Advance Praise for *Community Reviva*

"Having spent hundreds of hours interviewing New Orlea
math of Hurricane Katrina, my fellow researchers and I h
and scope of the challenges disaster victims must overcome homes
and their communities. In *Community Revival in the Wake of Disaster*, Storr, Haeffele-Balch, and Grube demonstrate vividly and compellingly how entrepreneurs across all sectors drive community recovery by providing necessary resources and coordinating recovery efforts. Scholars, students, and practitioners who are interested in how communities can rebound in the wake of disaster and how policymakers can promote resilient communities should read this book."

—Emily Chamlee-Wright, Provost and Dean, Washington College, USA; author of *The Cultural and Political Economy of Recovery: Social Learning in a Post-Disaster Environment*

"Too often we look only to the government to help with disaster recovery when, as this excellent book reveals, the answer is already right in front of us. Using hundreds of interviews and months in the field after Hurricane Katrina and Hurricane Sandy, the authors skillfully show us how entrepreneurs and local business people serve as change agents after crisis."

—Daniel P. Aldrich, Professor of Political Science, University Faculty Scholar, and Director of Asian Studies, Purdue University, USA; author of *Building Resilience: Social Capital in Post-Disaster Recovery*

"*Community Revival in the Wake of Disaster* focuses attention on the role that economic as well as social entrepreneurs may play in promoting community recovery and fostering resiliency. The case studies included in the book are compelling, and the broader lessons regarding the potential for social change in the wake of disaster are important."

—Lori Peek, Associate Professor of Sociology and Co-Director of the Center for Disaster and Risk Analysis, Colorado State University, USA; co-author of *Children of Katrina*

"Storr, Haeffele-Balch, and Grube take a critical and rejuvenating approach to the meaning of 'entrepreneur'. This book describes how post-disaster communities can make a comeback through collective return and renewal. As a retired fire marshal, I want this book in the hands of every community stakeholder. As an educator, I think it belongs in the hands of every student involved in becoming his or her community's future."

—Rodger E. Broomé, Assistant Professor, Department of Emergency Services, Utah Valley University, USA; Retired Battalion Chief and City Fire Marshal, West Jordan Fire Department, Utah, USA

"When disaster strikes a community, the natural instinct of those who want to assist in the recovery process is that 'we must do something,' which usually translates into a bureaucratic effort to centrally plan the recovery. However, in this book, the authors put forth a compelling case for decentralizing recovery efforts and allowing space for entrepreneurial activity to take place in the wake of a disaster. Storr, Haeffele-Balch, and Grube effectively argue that it is entrepreneurship that leads to a more robust and long-term recovery for the community affected by the disaster. This work is an important step in the process toward understanding the role that individuals and informal institutions play in post-disaster community recovery."

—Peter J. Boettke, University Professor of Economics and Philosophy, George Mason University, USA

PERSPECTIVES FROM SOCIAL ECONOMICS

Series Editor:
Mark D. White, Professor in the Department of Political Science, Economics, and Philosophy at the College of Staten Island/CUNY

The Perspectives from Social Economics series incorporates an explicit ethical component into contemporary economic discussion of important policy and social issues, drawing on the approaches used by social economists around the world. It also allows social economists to develop their own frameworks and paradigms by exploring the philosophy and methodology of social economics in relation to orthodox and other heterodox approaches to economics. By furthering these goals, this series will expose a wider readership to the scholarship produced by social economists, and thereby promote the more inclusive viewpoints, especially as they concern ethical analyses of economic issues and methods.

Published by Palgrave Macmillan

Accepting the Invisible Hand: Market-Based Approaches to Social-Economic Problems
Edited by Mark D. White

Consequences of Economic Downturn: Beyond the Usual Economics
Edited by Martha A. Starr

Alternative Perspectives of a Good Society
Edited by John Marangos

Exchange Entitlement Mapping: Theory and Evidence
By Aurélie Charles

Approximating Prudence: Aristotelian Practical Wisdom and Economic Models of Choice
By Andrew M. Yuengert

Freedom of Contract and Paternalism: Prospects and Limits of an Economic Approach
By Péter Cserne

Toward a Good Society in the Twenty-First Century: Principles and Policies
Edited by Nikolaos Karagiannis and John Marangos

Law and Social Economics: Essays in Ethical Values for Theory, Practice, and Policy
Edited by Mark D. White

Community Revival in the Wake of Disaster: Lessons in Local Entrepreneurship
By Virgil Henry Storr, Stefanie Haeffele-Balch, and Laura E. Grube

Community Revival in the Wake of Disaster

Lessons in Local Entrepreneurship

Virgil Henry Storr, Stefanie Haeffele-Balch, and Laura E. Grube

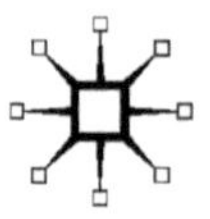

COMMUNITY REVIVAL IN THE WAKE OF DISASTER

First published in 2015 by
PALGRAVE MACMILLAN®

Palgrave Macmillan in the UK is an imprint of Macmillan Publishers Limited, registered in England, company number 785998, of Houndmills, Basingstoke, Hampshire, RG21 6XS.

Palgrave Macmillan in the US is a division of Nature America, Inc., One New York Plaza, Suite 4500, New York, NY 10004-1562.

Palgrave Macmillan is the global academic imprint of the above companies and has companies and representatives throughout the world.

ISBN 978-1-137-55971-5
E-PDF ISBN: 978–1–137–31489–5
DOI: 10.1007/978–1–137–31489–5

Library of Congress Cataloging-in-Publication Data

Storr, Virgil Henry, 1975–
Community revival in the wake of disaster : lessons in local entrepreneurship / Virgil Henry Storr, Stefanie Haeffele-Balch and Laura E. Grube.
pages cm.—(Perspectives from social economics)
Includes bibliographical references and index.

1. Entrepreneurship. 2. Social change. 3. Community development. I. Haeffele-Balch, Stefanie, 1984– II. Grube, Laura E., 1985– III. Title.

HB615.S767 2015
338′.04—dc23 2015014206

A catalogue record of the book is available from the British Library.

Contents

Figures and Tables

Figures

Tables

Preface and Acknowledgments

Like just about everyone in the United States, I spent August 29, 2005, and several days after, glued to my television set, flipping back and forth between cable news stations, stunned speechless as I watched the aftereffects of Hurricane Katrina in New Orleans. We were all watching what could only be described as the results of a perfect storm. A category three hurricane had hit a city that is 49 percent below sea level and where the levees neighboring some of the poorest areas were just not strong enough or tall enough to keep the Mississippi River in check. The city flooded. The streets became rivers. The rooftops became harbors. People stood on those rooftops, with spray painted messages like "please help us" and "the water is rising" on the shingles, hoping to be rescued by helicopter or boat. Like just about everyone in the United States, I cried that first night, overwhelmed by the pictures on my television screen as well as the stories of harrowing loss spattered with tales of individual heroism and resilience. If you had asked me in those days about the prospect of New Orleans recovering from Hurricane Katrina, I would have said that New Orleanians will probably eventually be fine but that New Orleans will never recover.

I didn't know then what I know now. I have spent hundreds of hours in the years following Hurricane Katrina talking to survivors and reading their recollections of their communities before the storm, their Katrina experience, and, importantly, why and how they rebuilt their homes and their communities despite the challenges. Since Katrina, I have also spent hundreds of hours reading, writing, and speaking about how communities rebound after disasters. I've become an expert on the subject. The marvel of disaster recovery, however, is that so many amateurs are able to figure it out. Individuals who have never suffered anything as catastrophic as a major hurricane or earthquake or flood are able to not only survive these disasters but also bounce back from them. Often, these amateurs do a better job than the experts in figuring out what their communities need to do to recover from disasters.

Disasters destroy property and wreck lives. Moreover, disasters are occurring at an increased rate and are more destructive now than they were in the past. Understanding the driving forces behind community recovery and how to foster resilient communities is, thus, more important now than ever.

Although there has been a lot written about how governments can promote recovery and solve the problems that occur after disasters, and there is now a sizable literature on the importance of social capital to post-disaster recovery, there has been relatively little written about the role of entrepreneurship in the post-disaster context. This is surprising because, as we argue, entrepreneurs are a key driver of post-disaster recovery.

Community Revival in the Wake of Disaster discusses how entrepreneurs help communities recover after disasters. We argue that entrepreneurs of all types, operating in market and nonmarket settings, in both the private and the public sector, recognize and pursue opportunities to promote community rebound after disasters. Specifically, their efforts make it more attractive and easier for disaster victims who were displaced by the disaster to return and rebuild their communities. This way of discussing the role of the entrepreneur in promoting post-disaster community recovery requires that we adopt an inclusive conception of the entrepreneur. This, of course, also has implications for public policy and, specifically, the strategies that we believe policymakers should advance if they wish their communities to be able to withstand and rebound from disasters.

Of course, *Community Revival in the Wake of Disaster* could not have been written without a lot of help. In fact, my coauthors and I owe thanks to a number of people who aided in the development of the project. I, for instance, continue to owe a huge debt to my mentor and dear friend, the late Don Lavoie. I have not written a word since becoming Don's research assistant in 1998 without wondering what Don would think about it. In fact, I worry more now about what he would think of my work than I did when I was a graduate student and I knew he would ask me to defend my arguments. Regarding this book, I believe that he would have appreciated the way we conceptualized the entrepreneur (chapter 2). Perhaps it is useful when theorizing to distinguish between commercial, social, political, and ideological entrepreneurs, but, in the real world, we argue, this obscures the similarities between the different types of entrepreneurs and overstates the differences. Don, an advocate of the adoption of qualitative methods for applied economic research, would have enjoyed the case studies of successful entrepreneurs in the post-disaster situation that we present in the book (chapters 5, 6, and 7). Admittedly, Don would have been critical of the inclusion of the model of the challenges associated with community recovery (chapter 3) as being too formalistic and he would have been disappointed in that our policy recommendations (chapter 8) leave so big a scope for government action. Overall, though, I think he would have liked the book.

Additionally, I owe a tremendous debt to two of Don's students: Peter J. Boettke and Emily Chamlee-Wright. Pete's vision led to the creation of the Mercatus Center's *Crisis and Response in the Wake of Hurricane Katrina* project and his guidance has been critical at every stage of that project, of which this book is a part. Emily invited me to be a coinvestigator with her on the project and to colead the teams that visited New Orleans and the greater region to interview victims of Katrina. Moreover, Emily has been my primary coauthor on the dozen plus shorter articles and book chapters that inform this book. Pete and Emily are also great friends. As I have said repeatedly, I could not wish for better siblings.

I would also like to thank my coauthors. I am proud to be able to call both Stefanie and Laura my coauthors and my friends. They have been true partners in this effort.

Special thanks is owed to my dear friend Paul Lewis who hosted me for a weeklong visit to the Department of Political Economy, King's College London, and organized a workshop on an earlier draft of the manuscript. The feedback that we received during that workshop has proven to be invaluable. And, an extra special thanks is owed to my friend Chris Coyne who looked at and offered detailed comments on multiple versions of the arguments in this book.

My coauthors and I would like to thank the trainers, leaders, and members of the various interview teams whose efforts have been critical to this project: Kate Linnenberg, Lenore Ealy, Daniel Rothschild, Nona Martin Storr, Mario Villarreal-Diaz, Anthony Skriba, Heather Allen, Ellenor O'Byrne, Erin Agemy, Katie Creel, Daniel D'Amico, Ian Hinsdale, Lorin Jones, Adam Martin, Kathleen O'Hearn, Brian Pitt, Marianne Rodriquez, Daniel Sacks, Emily Skarbek, Tyson Schritter, Andrew Serwadda, and Skyler Treat. We would additionally like to thank Khai Hoang for serving as an interpreter in the field and Vu Nguyen for translating interview transcripts.

My coauthors and I would also like to thank John Meadowcroft, Mark Pennington, Lynne Kiesling, Marc Sidwell, Emily Skarbek, David Skarbek, Steve Horwitz, James Witte, Susan Trencher, Paul Dragos Aligica, Solomon Stein, Rob Garnett, Claire Morgan, Peter Leeson, Fred Sautet, Arielle John, Bob Elder, and Nicola Virgill-Rolle for very helpful discussions and comments on earlier drafts of the arguments presented here. The usual caveat applies. We would also like to thank McKenzie Robey, Jessica Carges, and Lauren Thompson for their invaluable administrative support. Special thanks are owed to Mark White (series editor), Leila Campoli (editor), Sarah Lawrence (assistant editor), and everyone else at Palgrave Macmillan who helped to guide this book from development through to production.

We would like to thank Springer for allowing us to reuse portions of "The Capacity for Self-Governance and Post-Disaster Resiliency" (*Review of Austrian Economics*) that I coauthored with Laura E. Grube; Routledge for allowing us to reuse portions of "Post-Disaster Community Recovery in Heterogeneous, Loosely-Connected Communities" (*Review of Social Economy*) that I coauthored with Stefanie Haeffele-Balch; Wiley for allowing us to use portions of "Social Capital as Collective Narratives and Post-Disaster Community Recovery" (*The Sociological Review*); SAGE Publications for allowing us to use portions of "Club Goods and Post-Disaster Community Return" (*Rationality and Society*); and InderScience Publishers for allowing us to use portions of "The Role of Social Entrepreneurship in Post-Katrina Recovery" (*International Journal of Innovation and Regional Development*), all of which I coauthored with Emily Chamlee-Wright. We would like to thank the Mercatus Center at George Mason University for allowing us to reuse portions of "The Entrepreneur's Role in Post-Disaster Community Recovery: Implications for Post-Disaster Recovery Policy" and "Filling the Civil Society Vacuum: Post Disaster Policy and Community Response," which I coauthored with Emily Chamlee-Wright. And, we would like to thank Beloit College Press for allowing us to reuse portions of my articles, "The Determinants of Entrepreneurial Alertness and the Characteristics of Successful Entrepreneurs" that I coauthored with Arielle John, and "North's Underdeveloped Ideological Entrepreneur." We would like to thank De Gruyter for allowing us to reuse portions of "All We've Learnt: Colonial Teachings and Caribbean Underdevelopment" (*Le Journal des Economistes et des Etudes Humaines*).

This book could not have been written without the financial support of the Mercatus Center at George Mason University. Nor could it have been written had Brian Hooks and Daniel Rothschild not built and maintained a vibrant intellectual environment at Mercatus.

We would also like to thank our spouses (Nona, Alan, and Brandon) for their love, patience, and support while we developed this book. And, I would like to thank my daughter Winnie for being a perfect distraction and the best motivation since the day she was born.

Finally, we would like to thank the entrepreneurs who shared their stories with us. They taught us a great deal about commitment and courage, about hard work and heroism, and about resourcefulness and resilience. For that we will forever be grateful.

VIRGIL HENRY STORR
Manassas, Virginia
March 2015

Chapter 1

Introduction

Hurricanes, typhoons, earthquakes, tornadoes, and fires can shatter lives, destroy property, and cause severe emotional trauma. Consider, for instance, some of the worst disasters of the past few decades. The 2004 Great Sumatra earthquake and subsequent tsunami in the Indian Ocean affected almost a dozen countries, resulted in over 230,000 deaths, and displaced over 1.5 million people. It was the third largest and also the deadliest earthquake recorded in history, and destroyed over 300,000 homes, and resulted in over ten billion US dollars in damage. It is not hyperbole to state that many victims of this disaster lost everything; their homes, their loved ones, and their worldly possessions were swallowed up by the sea. The scale and scope of the devastation caused by this disaster was, in a word, overwhelming.

Similarly, the 2010 earthquake in Haiti caused over 160,000 deaths and left over 1.5 million Haitians homeless. Many people who did not have much to begin with lost even the little they had. Many of the homes in Port-au-Prince, prior to the earthquake, were shacks with dirt floors and corrugated roofs. For a decade before the earthquake, GDP per capita in Haiti hovered around $500 per year. The government in Haiti lacked the capacity to provide basic services for its citizens even before the disaster. It is difficult to imagine how this country could rebound from a disaster of this scale (i.e., one that cost as much as $13.2 billion in damage) without extensive assistance from outside the country.

Likewise, the disasters that we focus on in this book—Hurricane Katrina and Hurricane Sandy—caused significant harm. Hurricane Katrina and the floods that followed, for instance, caused over 1,800 deaths, resulted in over $100 billion in damage, and displaced over 400,000 Gulf Coast residents. Similarly, Hurricane Sandy caused over 180 deaths and resulted in over $60 billion in damage.

Although disasters can devastate communities, and the challenges associated with overcoming disasters would appear to make post-disaster community rebound unlikely, it is not uncommon for communities to eventually bounce back after disasters. Any post-disaster situation is certainly difficult and will seem nearly impossible to navigate, but communities around the world appear to be able to recover from disasters. How, then, do communities rebound after disasters?

Rebounding from a Disaster Is a Daunting Challenge

Community rebound after a disaster would seem to be a daunting challenge. Indeed, the costs associated with rebuilding after disasters are necessarily high. Rebuilding damaged or destroyed homes and businesses takes a great deal of time and a lot of money. Disaster victims have to buy building materials; hire builders; and replace lost clothing, furniture, appliances, and other household items. Rebuilding damaged or destroyed homes and businesses also takes a psychic toll on disaster victims. Picking through damaged keepsakes, recognizing that a space associated with key moments in one's life is now in ruin, and coming to terms with the sheer scale and scope of the loss can be extremely difficult. Moreover, the opportunity costs of rebuilding and not moving to start over elsewhere can be significant. The time spent attempting to "clear the rubble" where you lived prior to the disaster is time not spent building a life in a new community, seeking employment and building connections in that new place.

Often, community members lack the resources needed to rebuild a community after a disaster. Studies have found that the most economically and socially disadvantaged communities are also the most vulnerable to disaster.[1] Several studies have examined whether or not minority and low-income households have limited access to the resources that aid in preparing for, and recovering from, disasters. For instance, Peacock (2003) and Norris et al. (1999) found that minority and low-income households tend to be less prepared for hurricanes. Similarly, Blanchard-Boehm (1998) and Fothergill (2004) found that these households are less likely to have disaster-related insurance. Other studies have found that the poor suffer more injuries and deaths from disasters than the wealthy (see Wright 1979). Kahn (2005) found that developed countries, which have better-quality institutions, suffer fewer deaths after a natural disaster.

While the costs associated with community rebound after a disaster can be high and potentially prohibitive, the benefits associated with rebuilding are, at best, uncertain. In the wake of a disaster, every affected community member has to decide whether it makes more sense to move elsewhere or to attempt to repair or rebuild their damaged or destroyed homes and businesses. However, deciding whether or not to relocate in the wake of a disaster is a particularly difficult decision to make. One reason for the difficulty is that it is very hard for affected community members to predict what their community will look like in the months and years after the disaster. The questions that individuals must confront in mundane times are amplified after a disaster.

In order to assess the benefits of repairing and rebuilding damaged homes and businesses, disaster victims must find answers to many questions and scenarios. What will this community be like in the future? What business opportunities will exist? Similarly, business owners must figure out what their customer base is likely to be once displaced residents decide to return or relocate. Business owners must also figure out whether they will be able to find employees with the desired skillset and locate necessary suppliers before deciding to rebuild their businesses. Similarly, since businesses as well as homes are destroyed by disasters, affected community members must also figure out if they will find gainful employment after the storm. Additionally, displaced residents and business owners must form a judgment regarding which public services will be available in their community once recovery is under way. They must also make a judgment surrounding whether disrupted social networks will be restored, whether their churches and temples as well as social clubs will come back, and whether their neighbors, friends, and family members will rebuild rather than relocate.

These questions are virtually impossible to answer for several reasons. First, the post-disaster environment is unavoidably an environment in flux. Second, and perhaps most importantly, everyone's answers to these questions will depend in part on the decisions to relocate or rebuild of many others who are similarly struggling to find answers to the same questions before deciding what course of action to pursue. Another reason for the difficulty in deciding whether rebuilding or relocating is more advantageous is that the greater the extent of the disaster the more difficult it is for disaster victims to coordinate their activities with one another. Because disasters often displace community members, it can be difficult, if not impossible, to locate and reconnect with customers and employees; with friends, neighbors, and fellow churchgoers; and even with family members in the days, weeks, and months following a disaster. Because disasters create such uncertainty, and because there is greater uncertainty with greater disasters, the victims and affected members often have similar difficulties formulating and articulating concrete plans.

Community recovery after a disaster, thus, constitutes a collective action problem (Chamlee-Wright 2010; Chamlee-Wright and Storr 2009a, 2010a).[2] Arguably, the most sensible choice for every affected community member to make after a disaster is to wait and see what others decide before committing to a course of action. Stated another way, the risks associated with being a first mover could very well be prohibitive, and so the dominant strategy for every community member would be to wait and see what others decide to do.[3] If community members do not find a way to overcome this collective action problem, then community rebound after a disaster is unlikely to occur.[4]

Given the scale and scope of disasters, there is, not surprisingly, a tendency to focus on the various ways governments can help communities prepare and plan for, survive, and ultimately recover from, disasters. Governments, it is believed, are well positioned to marshal the resources and to mobilize the manpower needed to rebuild devastated communities after disasters.

There are, in fact, several ways governments have attempted to help communities rebound after disasters. They have sent in personnel to search for and to assist with the care for victims. They have sent in individuals to help with the cleanup in the immediate aftermath of the disaster. They have provided temporary housing for disaster victims in the weeks and months following a disaster. They have offered funds to help disaster victims rebuild their homes or businesses. They have repaired and reopened damaged and destroyed schools, repaired and restored disrupted public utilities (e.g., water, sewage, and electricity), and provided public services (e.g., police and garbage collection). They have also attempted to develop comprehensive plans for post-disaster community rebound.

While it is true that there is a tendency to look toward governments for assistance after disasters, there have certainly been some issues associated with the government provision of aid after disasters. There is also a realization that governments alone cannot bring about post-disaster community rebound. Indeed, nongovernmental organizations as well as private donors and volunteers have greatly assisted disaster victims in the aftermath of disasters (Chamlee-Wright 2010; Chamlee-Wright and Storr 2009a, 2010a; Shaw and Goda 2004; Zakour and Gillespie 1998). Additionally, disaster victims often rely on their social networks for assistance following a disaster (Aldrich 2011a, 2011b, 2012; Chamlee-Wright 2008, 2010; Chamlee-Wright and Storr 2010a, 2011b; Hurlbert et al. 2000, 2001; Murphy 2007; Storr and Haeffele-Balch 2012). Social networks, just as in mundane times, can be an important source of financial and emotional support in post-disaster contexts. Entrepreneurs, we argue, also play a critical role in bringing about community recovery after disasters.

Entrepreneurs Are Agents of Social Change

We think of entrepreneurs as agents of social change. As such, our notion of who counts as an entrepreneur is quite inclusive, encompassing not only commercial entrepreneurs but also social, political, and ideological entrepreneurs. When we describe someone as an entrepreneur, we do not mean to suggest that the person is necessarily the decision agent in a commercial enterprise who notices and seeks to exploit profit opportunities by developing and offering goods and services to customers at a price. We also think of the individual who organizes a food drive in their neighborhood and the community leader who organizes community members to protest some societal injustice as entrepreneurs. We also think of the religious leader who cares for his flock as an entrepreneur. We think of the political and bureaucratic leaders who propose and pass legislation and who advocate and direct policy as entrepreneurs. And, we also think of the thought leaders, who convince community members to change their views on social issues, as entrepreneurs. So, we would describe both the Bill Gates who started Microsoft, a for-profit software company, and the Bill Gates who started the Gates Foundation, a nonprofit charitable enterprise, as entrepreneurs. We would describe industrialist Henry Ford, who revolutionized the production of automobiles, and President Woodrow Wilson, who pushed for a change in the ways nations interacted with one another, as entrepreneurs.

Although our conception of who counts as an entrepreneur is inclusive, and in some sense almost everyone can be described as acting entrepreneurially at various moments in their lives, it does have some boundaries.[5] Entrepreneurs qua entrepreneurs notice opportunities (whether writ large or writ small) and move to exploit them to change the world. The functionary who is implementing the entrepreneur's vision is not an entrepreneur. Or, rather, he is not acting as an entrepreneur when merely carrying out someone else's vision to change the world. Similarly, the financier or donor who is merely bankrolling the operation is not acting as an entrepreneur. Or, rather, she is not acting as an entrepreneur when merely providing the resources to support someone else's vision to bring about social change. Likewise, the recipient or the customer who is benefiting as a consequence of the entrepreneur working to implement her vision is not acting as an entrepreneur. Or, more specifically, she is not acting as an entrepreneur when merely securing or enjoying the fruits of an entrepreneur's efforts.

When conceived this way, it is easy to imagine the critical role entrepreneurs play in societies in mundane times. After a disaster, entrepreneurs, we argue and hope to show, play an important role in helping communities overcome the collective action problem that characterizes community rebound. Specifically, we focus on their post-disaster efforts to provide necessary goods and services, restore and replace disrupted social networks, and signal that community rebound is likely and, in fact, under way. Each of these actions reduces the costs associated with rebuilding a damaged or destroyed home or business, increases the benefits associated with post-disaster community rebound, or helps to convince disaster victims that others also believe that the benefits associated with rebuilding outweigh the costs.

Bureaucrats, on the Other Hand, Should Be Seen as Promoting Stability

Entrepreneurs are creative, innovative figures who utilize their positions, experiences, and resources to identify opportunities and drive social change. Bureaucrats, on the other hand, are rule-following figures that sustain rather than challenge the status quo. The policies bureaucrats propose and the strategies they pursue are likely to be characterized by balance and predictability. Rather than being disruptive figures whose actions bring about change, bureaucrats are likely to adopt routines, advocate caution, and resist change.[6]

When we describe bureaucrats, it is important to note that we do not mean to suggest they occupy a particular position within an organizational chart, nor do we mean to suggest they are a part of any particular kind of organization. Instead, in describing someone as a bureaucrat, we are ascribing to that particular person in that particular moment a particular mind-set and suggesting that she is performing a particular social function. Bureaucrats, thus, can exist in both the public and private sectors, in both for-profit and nonprofit organizations, and at both the top and bottom of a given organizational chart. Whereas entrepreneurs notice opportunities in all environments and move to exploit them for social change, bureaucrats can be seen as implementing plans, following rules, and promoting stability.

Different types of environments (e.g., monocentric versus polycentric systems), however, can either encourage and give greater scope for entrepreneurial efforts to bring about social change or can promote bureaucratic efforts to preserve the status quo. Monocentric systems are systems where power is centralized. Polycentric systems have multiple centers of power. Monocentric systems are typically characterized by strict protocols and chains of command and tend to limit innovation in favor of maintaining the status quo. Polycentric systems, however, give individuals the space to experiment and tend to allow for autonomy and encourage challenges to the status quo. Whereas monocentric systems tend to be rigid and hierarchical, polycentric systems are more decentralized. Bureaucrats are closely associated with monocentric systems, and entrepreneurs are more likely to thrive in polycentric systems.

Although the differences between a monocentric system and a polycentric system are quite stark and are likely to be readily apparent to anyone operating within them, it is difficult to determine whether an organization promotes entrepreneurship or bureaucratic behavior by studying its organizational chart. For instance, two organizations might have similar formal hierarchies—say, they each have a president, senior managers, middle managers, and staff—but one may encourage entrepreneurship at various levels within the organization and the other might discourage it. Instead of looking at the organizational chart, it is important to examine the actual characteristics of and dynamics within the organization. In other words, examining the rules, roles, and expectations that in fact govern interaction within an organization is key to understanding the nature of that organization. Stated another way, it is the rules-in-use and not the rules-in-form that ultimately matter. For instance, the owner of a small company with a few staff members and a flat organizational structure may, nonetheless, tightly control the organization and give little scope for employees to act outside their prescribed roles. While the owner may be acting entrepreneurially in bringing new products to the market, he may close off the possibility of his employees acting entrepreneurially within and on behalf of his firm. Likewise, a large company or a federal agency may have formal structures that appear quite hierarchal but might also give middle managers autonomy to make decisions on the ground.

When the challenges confronting a community are complex and the circumstances are uncertain and constantly changing, such as during and after a disaster, there arguably needs to be an even greater scope for entrepreneurial efforts than in mundane times. Moreover, if too many people adopt a bureaucratic mind-set within government agencies tasked with assisting communities and within private organizations affected by disaster, they might actually retard rather than promote recovery. Stated another way, polycentric rather than monocentric systems are likely to be better able to bring about community rebound after a disaster. This is, of course, not to say that during and after a disaster there is no role for bureaucrats who simply follow orders and work to accomplish the tasks put before them. It is difficult, for instance, to imagine any medium- or large-scale operation that could be effective if all the people taking part in it were simply told to look for opportunities to promote change and pursue them. Instead, pointing to the importance of entrepreneurship after disasters highlights that, given the challenges associated with post-disaster community recovery, bureaucrats alone cannot bring about recovery, and that failing to give entrepreneurs enough scope to act will hamper community recovery.

And, so, This Book Argues That...

Entrepreneurship is a key force behind community rebound after disasters. After a disaster, entrepreneurs, conceived broadly as individuals who recognize and act on opportunities to promote social change, play a critical role in helping communities overcome the collective action problem that characterizes post-disaster community rebound. Specifically, we argue that entrepreneurs promote community recovery after disasters by providing necessary goods and services, restoring and replacing disrupted social networks, and signaling that community rebound is likely and, in fact, under way. Given the importance of entrepreneurship after disasters, we argue that policymakers should recognize that bureaucrats alone cannot bring about recovery and should attempt to create environments where entrepreneurs have the space to act.

This project should, thus, appeal to (1) those interested in disaster recovery and community rebound as well as (2) those interested in how coordination problems are solved in the real world. The first group of scholars, policymakers, and interested nonspecialists (i.e., those interested in disaster recovery and community rebound) tend to be concerned with how communities are impacted by disasters, how communities can recover from disasters, and what can be done to make communities more resilient. Arguably, the members of this group tend to emphasize centralized disaster management approaches and tend to discount the role of decentralized efforts in helping to solve the massive problems that affect entire communities.[7] Our project, however, suggests that this group overemphasizes the importance of monocentric approaches to bring about community recovery and underemphasizes the role of entrepreneurship after disasters.

The second group of scholars, policymakers, and interested nonspecialists tend to emphasize the role of entrepreneurs in overcoming real-world coordination problems. Not surprisingly, members of this group tend to argue that commercial entrepreneurs are better positioned than entrepreneurs in nonmarket settings and certainly better positioned than bureaucrats in government to solve complex coordination problems.[8] Our project suggests that this group underappreciates the potential of nonmarket entrepreneurship—such as social, ideological, and political entrepreneurship—to solve coordination problems and bring about social change.

Community Revival in the Wake of Disaster is organized in nine chapters. Chapter 2, "The Entrepreneur as a Driver of Social Change," begins with a discussion of Israel Kirzner's and Joseph Schumpeter's seminal theories of entrepreneurship in the commercial sphere. While Kirzner stresses that entrepreneurs notice profit opportunities, Schumpeter emphasizes the role of entrepreneurs in introducing new goods and services and developing new ways of producing or delivering existing goods and services. We argue that their theories of commercial entrepreneurship can be extended to explain social, political, and ideological entrepreneurship. We further argue that there are important similarities between these different kinds of entrepreneurship and that we can conceive of entrepreneurs more generally as social change agents.

Chapter 3, "How Entrepreneurs Promote Post-Disaster Community Rebound," outlines the challenges associated with recovering from a disaster and highlights several key ways by which entrepreneurs help bring about post-disaster community rebound. Specifically, we focus on entrepreneurs providing needed goods and services, restoring and replacing disrupted social networks, and signaling that community rebound is likely and, in fact, under way.

Chapter 4, "How Entrepreneurship Promotes Community Recovery: The Cases of Hurricanes Katrina and Sandy," describes our empirical approach and offers further details about the challenge of rebounding from and the extent of the damage caused by Hurricane Katrina and Hurricane Sandy.

Next, chapters 5, 6, and 7 offer specific examples of entrepreneurs promoting community rebound after these disasters. In chapter 5, "Goods and Services Providers," we argue that entrepreneurs provide needed goods and services during mundane times, and that it is reasonable to think they may also provide key goods and services in post-disaster scenarios. As we discuss this topic, a growing literature does in fact support the theory that entrepreneurs may effectively provide goods and services in hostile environments (Bullough et al. 2013; Galbraith and Stiles 2006). Following a disaster, victims often require food, water, and shelter. They may also need clothing, day care services, or transportation. As rebuilding begins, the goods and services demanded may shift to items related to construction, appliances, and furniture, as well as the specialized skills offered by architects, plumbers, and electricians. Indeed, entrepreneurs provide all these things and more. For instance, we examine how one entrepreneur started a health clinic in the Lower Ninth Ward after Hurricane Katrina and how the leader of a civilian patrol in Bayswater and Far Rockaway used his existing group to help ensure community safety in the aftermath of Hurricane Sandy.

In chapter 6, "Regrowing Uprooted Social Networks," we focus on the entrepreneur as a restorer of social networks in a community. Entrepreneurs can reach out to investors and donors, employees, and volunteers, as well as customers and clients who may be dislocated as a result of the disaster, and they can communicate to others information about available resources and may be able to obtain resources for disaster victims. Again, we study several entrepreneurs who have helped to restore and replace social networks in their communities following a disaster. We illustrate, for example, how the leader of a local community organization in New Orleans used the organization's membership records to reconnect with residents who had evacuated and not yet returned, as well as how a resident of Bayswater used his own connections within the community to engage with a larger social network, including those who wanted to provide disaster assistance and those who needed such assistance.

In chapter 7, "Entrepreneurs as Signals of Healthy Community Rebound," we describe how entrepreneurs can coordinate recovery efforts and serve as important focal points. For instance, we explore how a pastor of the Mary Queen Vietnam Catholic Church took deliberate steps to signal members to return to his community, and how the owner of a convenience store in the Lower Ninth Ward worked hard to get his business back up and running and, in so doing, convinced community members in the surrounding neighborhood that recovery was under way.

In chapter 8, "Fostering Resilient Communities," we argue that giving entrepreneurs the space to act is crucial for fostering resilient communities. The work of Vincent and Elinor Ostrom provides a useful distinction between monocentric systems, which are characterized by strict hierarchy and tend to promote the status quo, and polycentric systems, which are characterized by decentralized and competing nodes of authority and tend to promote entrepreneurial activity. Since entrepreneurs play a crucial role in promoting community rebound after disasters, we argue that policymakers should create polycentric systems as opposed to monocentric ones for disaster management, and they should encourage entrepreneurial efforts in both the public and private sphere to bring about post-disaster community rebound. We do not argue, however, that the government does not have a role to play in helping communities to recover after disasters. In fact, it is possible to imagine policymakers accepting our policy recommendations and still pursuing a sizeable government response to disasters. Our recommendations do not argue against government responses to disasters per se, but we do argue against government responses to disasters that limit the scope of post-disaster entrepreneurship. Our recommendations speak against structuring government responses in a way that does not allow for experimentation, innovation, and decentralized responses to disasters.

Finally, in chapter 9, "Conclusion," we offer concluding remarks. In summary, the entrepreneur is a driver of social change and is a driving force behind post-disaster community rebound. Specifically, entrepreneurs provide needed goods and services, help community members restore or replace disrupted social networks, and signal that community rebound is likely and, in fact, under way. Since entrepreneurs play such a critical role in promoting post-disaster community rebound, policymakers should ensure that entrepreneurs have the space to act.

Chapter 2

The Entrepreneur as a Driver of Social Change

Every good plot needs a protagonist. Of course, the setting where the protagonist's story plays out, the constraints they must work within, and the circumstances that the characters must confront all matter. It matters, for instance, that Shakespeare's Henry V is set in fifteenth-century England and France, that Henry V is a newly crowned King of England who also has a claim to the French Crown, and that the English army is much smaller than the French army they must combat. But without Henry V there is no British invasion of France, there is no Battle of Agincourt, and there is no British monarch on the French throne. In short, without Henry V there is no play.

Similarly, the entrepreneur is a central figure in most episodes of social change. Of course, the legal environment in which entrepreneurs must operate, the budget constraints that delimit their options, and the social ties and other factors they must consider before they can choose a course of action, all shape entrepreneurial activity. But it is impossible to imagine new products being introduced to the market, new policies being implemented by government agencies, or new social programs being developed without an entrepreneur recognizing the possibility to do it and initiating the effort. In other words, the entrepreneur is often the proximate cause of social change.

As we argue below, focusing on the essential role that the entrepreneur plays in recognizing and exploiting opportunities to promote social transformation is essential to understanding the essence of entrepreneurship as well as the dynamics of social change.

Kirzner and Schumpeter Have Taught Us a Lot about Entrepreneurs

While much has been written on the differences between the entrepreneurial theories developed by Israel Kirzner and Joseph Schumpeter, we argue that they offer broadly complementary views of the entrepreneur. Kirzner and Schumpeter, simply, emphasize different moments of the entrepreneurial process. While Kirzner stresses the entrepreneur's alertness to opportunities for gain, Schumpeter stresses the characteristics she tends to exhibit and the outcome of her efforts. Stated another way, Kirzner focuses on opportunity identification and Schumpeter focuses on opportunity exploitation (Kirzner 1999). Or, as Kirzner and Sautet (2006) explain, Schumpeter focuses on the behavioral aspects of entrepreneurship (i.e., boldness, creativity, and innovation) and Kirzner focuses on the cognitive aspects of entrepreneurship (i.e., opportunity identification, recognition, and discovery).

Kirzner Stresses the Cognitive Aspects of Entrepreneurship

For Kirzner, the entrepreneur is at center stage in the market process and, so, in the process of economic development. As Kirzner (1997: 31) summarizes, there is "a powerful tendency [within market economies] for less efficient, less imaginative courses of action, to be replaced by newly discovered, superior ways of serving consumers—by producing better goods and/or by taking advantage of hitherto unknown, but available, sources of resource supply." Entrepreneurs in the market, Kirzner explains, notice and exploit opportunities to earn profits by replacing "less efficient, less imaginative courses of action" with "superior ways of serving consumers." For Kirzner, the entrepreneur also plays a key but underappreciated role in microeconomic theory.

Ironically, though most economists will employ the entrepreneur in classroom discussions of how the market moves from disequilibrium to equilibrium, "standard theory has not been able to explain how markets systematically gravitate towards the equilibrium states (relevant to the given conditions of those markets)" (Kirzner 2009: 147). Introducing the entrepreneur and outlining her role in "driving the process of equilibration," according to Kirzner, is critical to understanding the market as a dynamic process (ibid.: 147). Kirzner's theory of entrepreneurship, thus, makes an important contribution to standard price theory and our understanding of real-world markets.

In an effort to understand the nature of the market process and to tease out the essential role that the entrepreneur performs, Kirzner (ibid.: 147) begins his theorizing in the "*simplest* Marshallian demand-supply context," that is, a single period world with a single commodity and no scope for uncertainty. Kirzner relaxes only one assumption within this framework (i.e., the "perfect knowledge" assumption). In a world of perfect knowledge, individuals do not really make decisions about how to succeed but instead calculate optimum strategies on the basis of given data. Deciding is merely a matter of deciding to exchange something that is less preferred or valued for something that is more preferred or valued. The result of this "decision" is implied in and perfectly determined by the decision maker's situation. "If each individual knows with certainty what to expect," as Kirzner (1973: 37) explains, "his plans can be completely explained in terms of economizing, of optimal allocation, and of maximizing—in other words, his plans can be shown to be in principle implicit in the data which constitutes his knowledge of all the present and future circumstances relevant to his situation." In such a world, there is no possibility of the same good selling for different prices. Any apparent price discrepancies that might exist would necessarily be explained by differences in transaction costs, like differences in the transportation costs between the space where the good was produced and the two points of sale. As such, in a world of perfect competition, there is no possibility of earning pure profits (that is, profits apart from normal rates of return on investments) and so there is simply nothing for the entrepreneur as a seeker of pure profit to do.

In Kirzner's view of the world, however, knowledge is imperfect and cannot be perfected. As Kirzner recognizes, there is never a time in the real world when all possible means have already been identified and all possible ends have already been conceived. Instead, human beings live in a world of radical uncertainty and fundamental ignorance (Mises [1949] 1963; Hayek 1948, 1955; Kirzner 1973, 1979).[1] It is because of this uncertainty that the entrepreneur has a function and that the market process takes place.

According to Kirzner, there is much "out there," as it were, waiting to be discovered.[2] For instance, sellers are often unable to locate prospective buyers for their goods and services. Similarly, buyers in one area are often unaware that a good or service that they desire is available in a neighboring locality. And, because knowledge is imperfect, Kirzner (1973) points out, there is the possibility of the same good selling for different prices in the market and, thus, the possibility for arbitrage. As an example, in a world of imperfect knowledge, Store A may be selling a particular Brand X tennis shoe for $50 while Store B sells the same Brand X tennis shoe for $100, with neither the storeowners nor their customers being aware of the price discrepancy. In such a market, it would be possible for some individuals to buy tennis shoes from Store A and sell them to the customers of Store B at a profit.

These arbitrage opportunities exist because, where knowledge is imperfect, buyers and sellers can make errors of overoptimism, which lead to frustrated plans as well as errors of overpessimism, which lead to unexploited opportunities. Because of their ignorance and "errors made in the course of market exchanges," Kirzner (1999: 6) explains, market participants can be led

> (1) overoptimistically to insist on receiving prices that are "too high" (to enable them to sell all that they would like to sell at those prices) [or on paying prices that are 'too low' (to enable them to buy all that they would like to buy at those prices)]; or (2) over-pessimistically to enter into transactions that turn out to be less than optimal in the light of the true market conditions as they in fact reveal themselves (e.g., a buyer discovers that he has paid a price higher than that being charged elsewhere in the market; a seller discovers that he has accepted a price lower than that which has been paid elsewhere in the market).[3]

The overoptimistic seller is unable to sell her wares. Similarly, the overoptimistic buyer is unable to find the goods she desires at the price she is willing to pay. On the other hand, the overpessimistic buyer or seller has left money on the table. She could have made more or saved more than she did on the transaction.

Kirzner's entrepreneurs are *alert to* these arbitrage opportunities (i.e., they look for and recognize these opportunities to buy low in one market and sell high in another). By alertness, Kirzner (1994: 107) means a "generalized intentness upon noticing the useful opportunities that may be present within one's field of vision," that is, an intentness on noticing unexploited profit opportunities. This includes recognizing that a set of raw materials, machinery, and labor can be bought/rented in factor markets and combined to make a product that can be sold at a profit in consumer markets. Kirzner (1994: 107) has described alertness as his entrepreneur's chief characteristic. The entrepreneur, thus, plays an essential role in getting buyers the goods and services that they want.

According to Kirzner, it is their alertness to these opportunities that explains the tendency of entrepreneurs to equilibrate the market.[4] As Kirzner (2000: 16) explains, "each market is characterized by opportunities for pure entrepreneurial profit. These opportunities are created by earlier entrepreneurial errors which have resulted in shortages, surplus, and/or misallocated resources. The daring, alert entrepreneur discovers these earlier errors, buys where prices are 'too low' and sells where prices are 'too high.'" If these errors/opportunities are to be discovered, individuals must be vigilant and exhibit a "natural alertness" to such errors (ibid.: 18). As Kirzner (ibid.: 23) writes, alertness is a key factor, "discovery is attributable, at least in significant degree, to the entrepreneurial alertness of the discoverer."

This alertness, it is important to point out, is quite different from possessing superior knowledge. It is true that the entrepreneur knows something that his fellow market participants do not know. He knows, for instance, that there is an opportunity to buy Brand X tennis shoes from Store A and sell it at a higher price to the customers of Store B. But it is not that he possesses superior knowledge. He is not at all like the doctor who possesses superior knowledge of medicine than lay individuals. Instead, the essence of entrepreneurship is that the entrepreneur notices an opportunity when others who might have noticed it did not.[5]

It is also important to point out that deliberately searching for profit opportunities is altogether different than an entrepreneur being alert to and discovering profit opportunities. An individual will only decide to engage in a deliberate search for knowledge if the anticipated costs of searching are less than the expected returns from searching.[6] Consequently, any so-called search for profitable opportunities necessarily follows the entrepreneur noticing that there is an opportunity to potentially benefit from a particular kind of search.[7]

For Kirzner, then, entrepreneurship is an equilibrating force that consists of (and is reducible to) an alertness to arbitrage opportunities that are readily discernable and that only exist because of widespread ignorance in the market. Admittedly, this view of the market process is quite simplistic. In the real world (i.e., where there are multiple periods and multiple commodities), market participants are necessarily uncertain as to the best course of action available to them (i.e., there is scope for creativity and judgment). Furthermore, they are not able to unambiguously make sense of their circumstances and their options (i.e., there is scope for interpretation). However, in working with this simple model, Kirzner is able to isolate the essence of entrepreneurship. He is able to demonstrate that if the market is to move toward equilibrium (i.e., the elimination of errors of overoptimism and overpessimism), it must be comprised of not only agents who can economize (i.e., Robbinsian maximizers) but must also contain agents who are alert to profit opportunities (i.e., pure entrepreneurs). Kirzner is also able to isolate a key moment of entrepreneurship (i.e., opportunity identification).

Kirzner's theory of entrepreneurship has had considerable influence in the fields of economics and entrepreneurship studies.[8] Gaglio and Katz (2001), for instance, agree with Kirzner that opportunity identification is the most fundamental aspect of entrepreneurial action. Many scholars have utilized the Kirznerian theory of opportunity recognition to advance contemporary entrepreneurship research (Gartner 1990; Shane and Venkataraman 2000; Shane 2003; Dimov 2007). Indeed, Klein and Bylund (2014: 262) examine citation patterns and find that Kirzner "ranks as one of the most highly cited economists in the management and entrepreneurship literature, behind Schumpeter and Oliver Williamson." Minniti and Koppl (2003) have also highlighted the contribution of Kirzner's theory to entrepreneurship studies. See also Foss (1997), Endres and Woods (2006), Hébert and Link (1989), Klein (2012), and Shane (2000).

Others have sought to extend or apply Kirzner's conception of entrepreneurship. Arentz et al. (2013), for instance, conducted experiments that showcase the importance of prior knowledge in opportunity identification. Similarly, Yu (2001) focuses on opportunity identification and the subjective nature of alertness, emphasizing how the discovery process is shaped by one's stock of knowledge. He then advances two types of discovery, the ordinary (which are backward-looking and aim to improve existing situations) and the extraordinary (which are forward-looking innovations). Additionally, Denrell et al. (2003) and Zaheer and Zaheer (1997) apply the concept of alertness to firms. Other research has focused on distinguishing between alertness and other cognitive functions, or personality traits, needed to recognize and carry out entrepreneurial endeavors (see, for instance, Ray and Cardozo 1996; Baron 1999; Keh et al. 2002; Gaglio 2004; Moreno 2007).

Furthermore, Kirznerian entrepreneurship has been utilized in organizational theory, development economics, economic history, and other topics of study. For instance, Sautet (2000) employs the entrepreneur in a process-oriented theory of the firm. Storr (2004, 2006, 2013) deploys Kirzner's theory of entrepreneurship to examine the entrepreneurial narratives in Bahamian culture. Further, Tominc and Rebernik (2007) and Runst (2013) utilize Kirznerian theory to examine entrepreneurship in Eastern Europe and Germany, respectively. Kirzner's theory of entrepreneurship has also been used to examine economic development and institutional change to highlight how different institutional arrangements alter the opportunities that entrepreneurs are alert to and the activities they pursue (Boettke and Coyne 2003, 2009; Leeson and Boettke 2009; Sautet 2008). Bjørnskov and Foss (2008), for instance, examine the relationship between economic freedom (as measured by the Economic Freedom of the World Index) and the supply and allocation of entrepreneurship, and they find that lower economic freedom (or more government intervention) distorts price signals and, therefore, adversely affects entrepreneurial activity.

Schumpeter Stresses the Behavioral Aspects of Entrepreneurship

Schumpeter has characterized the entrepreneur as the sine qua non of the development process. Without the entrepreneur, economic development, as Schumpeter understands it, cannot occur. For Schumpeter ([1934] 2012: 66), development is "defined by the carrying out of new combinations."[9] Development, for Schumpeter (ibid.: 68), "consists primarily in employing existing resources in a different way, in doing new things with them, irrespective of whether those resources increase or not." As such, development is a process of discontinuous change in the economic system that appears "in the sphere of industrial and commercial life" (ibid.: 65).[10] It is a process of creative destruction where, through the "carrying out of new combinations," the economic system is consistently revolutionized. Specifically, Schumpeter (ibid.: 66) sees development as consisting of the following:

> (1) The introduction of a new good—that is one with which consumers are not yet familiar—or of a new quality of a good. (2) The introduction of a new method of production, that is one not yet tested by experience in the branch of manufacture concerned, which need by no means be founded upon a discovery scientifically new, and can also exist in a new way of handling a commodity commercially. (3) The opening of a new market, that is a market into which the particular branch of manufacture of the country in question has not previously entered, whether or not this market has existed before. (4) The conquest of a new source of supply of raw materials or half-manufactured goods, again irrespective of whether this source already exists or whether it has first to be created.[11]

Schumpeter is careful to point out that economic development is not merely the discovery of resources that had previously been idle. Nor is development simply a matter of discovering more untapped resources. Development is also not just the result of the natural increase in the supply of productive means that occurs as a result of population growth or savings and investment. These certainly contribute to development, but they are not the central force behind economic development. Instead, Schumpeterian development consists primarily in "the different employment of the economic system's existing supplies of productive means" (ibid.: 68). This process, however, does not just happen but needs a catalyst, someone or some agent-type, to set it in motion.

Schumpeter sees the entrepreneur as the star of this drama, the catalyst that sets the process of development in motion. His entrepreneur is an innovator who initiates new enterprises (i.e., "the carrying out of new combinations") and, as a result, extends the range of possible profit opportunities. She has an incentive to do this, to initiate new enterprises, because of the possibility of capturing entrepreneurial profits and, according to Schumpeter, she has the capacity to do this because of the availability of credit.[12] As Schumpeter ([1942] 1976: 132) writes, "the function of the entrepreneur is to reform or revolutionize the pattern of production by exploiting an invention." Because the entrepreneur is a pioneer who challenges conventions (economic and perhaps even social), Schumpeter's entrepreneur is a bold and creative visionary. He must act confidently as he ventures into unknown territory. The entrepreneur's role, Schumpeter (ibid.: 133) writes, "though less glamorous than that of medieval warlords, great and small, also is . . . just another form of individual leadership acting by virtue of personal force and personal responsibility for success." Rather than being motivated primarily by a quest for glory and political power like the medieval warlord, however, Schumpeter's entrepreneur is primarily motivated by a desire for profits and economic power.

Schumpeter's theory of entrepreneurship has had considerable influence within the fields of organizational theory, economic history, and entrepreneurship studies (see Bauer 1997; Breschi et al. 2000; Brouwer 2000; Giersch 1984; Perelman 1995; Wiggins and Ruefli 2005). Giersch (1984) presents an argument on how Schumpeter has influenced economics and thought within the social sciences more generally. According to Giersch (ibid.), Schumpeter pushed back aspects of Keynesian economics, reemphasized methodological individualism (and therefore, microeconomics) and moved away from steady state equilibrium and toward business cycles and innovation. Perelman (1995) also recognizes Schumpeter's contribution to economic thinking and links his work to an earlier economist, David Wells. Wiggins and Ruefli (2005) consider Schumpeter's version of economic history and his claim that economic advantages will become more and more difficult to sustain, connecting his theory to more recent literature on hypercompetition. They show, for example, that periods of sustained advantage have been growing shorter in the US economy.

Others have sought to apply or extend Schumpeter's theory of entrepreneurship. Brouwer (2000), for instance, seeks to combine Schumpeter's and Knight's theories of entrepreneurship to develop a theory of how innovation is affected by the number of incumbent competitors and by the number of new entries. Additionally, Breschi et al. (2000) use Schumpeter's theory of innovation to distinguish between two patterns of innovative activities, creative destruction (widening) and creative accumulation (deepening), and identify factors that lead to those patterns. Furthermore, Bauer (1997) engages Schumpeter's theories of innovation and creative destruction to analyze the implications of telecommunication regulations.

There is also a body of literature that explores Schumpeter's investigation of innovation and market structure, including characteristics of market sellers, barriers to entry, and firm size (Baldwin and Scott 1987; Kraft 1989; Maclaurin 1954). Baldwin and Scott (1987), for instance, examine Schumpeter's claim regarding the importance of firm size and market power in a firm's ability to compete in the modern industrialized world, noting that investment in research and development allows large firms to develop new products and processes. Similarly, Kraft (1989) provides an empirical analysis of the metal industry in West Germany, and confirms Schumpeter's emphasis on the importance of larger firm size and access to internal financing for innovation. Furthermore, relying on Schumpeter's theory of development, Maclaurin (1954) looks at 13 American industries and finds that industries that are close to the consumer, such as automobiles, food, and paper, internally invest in research and development, whereas more removed industries, such as chemicals, receive research funding through government contracts. He notes that contracts can nudge innovation in particular directions, whereas internally funded research may be driven by efforts to find new products and processes that consumers will demand.

There Are Broad Similarities between Kirznerian and Schumpeterian Entrepreneurs

There is a longstanding debate among economists on whether Schumpeter's and Kirzner's theories of entrepreneurship can be reconciled (Boehm 1990; Boudreaux 1994; Choi 1995; Hébert and Link 1982). Kirzner (1999), for instance, has suggested that his theory of entrepreneurship and Schumpeter's conception should be viewed as complements rather than alternatives. As Kirzner (1999: 12) explains, "Once we permit the multi-period character of real world entrepreneurial behavior to be explicitly considered, the relevance of the active aggressive characteristics of Schumpeter's entrepreneurs becomes understandable and important. Entrepreneurial alertness, in this essentially uncertain, open-ended, multi-period world must unavoidably express itself in the qualities of boldness, self-confidence, creativity and innovative ability." Stated another way, Schumpeter and Kirzner are dealing with two different moments of entrepreneurship. And, entrepreneurs in the real world must not only notice opportunities but must also possess certain qualities in order to exploit them.

In addition to their emphasizing different moments of the entrepreneurial process, another key difference between Schumpeter's and Kirzner's theories is how they describe entrepreneurial activity and its relationship to market equilibrium. As noted above, Schumpeter ([1942] 1976) sees entrepreneurs as disrupting the market processes by changing the methods of production, producing a new commodity, or discovering new inputs to produce an existing product. These activities are necessarily disequilibrating. Kirzner's entrepreneur, in contrast, is alert to profit opportunities (i.e., errors of overoptimism or overpessimism that result from uncertainty) and acts to push the market toward equilibrium.

Boudreaux (1994), however, argues that this difference between Kirzner and Schumpeter is more apparent than real. As Boudreaux (ibid.) suggests, the equilibrating/disequilibrating entrepreneur debate arises from a less than satisfactory definition of equilibrium. By defining competition as moving from a less desirable price-quantity combination to a more desirable price-quantity combination, both conceptions of the entrepreneur may be considered as equilibrating. Choi (1995) also acknowledges that Schumpeter and Kirzner handle equilibrium theory in two very different ways but argues that, at their core, both do agree on several key points. Specifically, Choi (ibid.: 62) argues that Schumpeter and Kirzner agree on three points, "(1) that there exist in the economy unexploited opportunities, (2) that the role of the entrepreneur consists of exploiting them, and (3) that traditional economic theory is flawed in leaving out the existence of unexploited opportunities and thereby overlooking the very force that moves the economy."

There are, indeed, broad similarities between the two figures. Both are driven by profit opportunities. Both also succeed by creating new products or markets. While Kirzner characterizes this as discovery and Schumpeter describes this as creative destruction, both view the entrepreneur as seeing some possibility (e.g., to profitably combine factors in a particular way) that no one else has identified or attempted to pursue. Both Schumpeter and Kirzner conceive of the entrepreneur as someone who must be willing to be a believer in a world of skeptics.[13]

Scholars Have Discussed Entrepreneurship in Nonmarket Settings

Arguably, any general theory of entrepreneurship, any theory of entrepreneurship that discusses entrepreneurship in commercial as well as noncommercial settings, would have to draw on both cognitive and behavioral approaches to conceptualizing entrepreneurship. Although Schumpeter focuses on the "carrying out of new combinations" in the commercial sphere, individuals certainly attempt to innovate in noncommercial settings. Similarly, Kirzner (1973) sees entrepreneurship as an element of all human action, be it economic, political, or social action.[14]

Not surprisingly, Kirzner's and Schumpeter's conceptions of the entrepreneur have been profitably used to discuss entrepreneurship in markets as well as nonmarket settings, such as political entrepreneurship. Shockley et al. (2006), for instance, combine Buchanan and Tullock's (1999) public choice insights with Kirzner's and Schumpeter's theories of entrepreneurship to build a theory of "public sector entrepreneurship."[15] Public sector entrepreneurs, they explain, are politicians or officials who are alert to, and work to, exploit opportunities to earn political profits (i.e., political and bureaucratic power). Similarly, extending Kirzner's formulation, Holcombe (2002) describes political entrepreneurs as individuals who observe and act on the predatory entrepreneurial opportunities that exist in government. Wagner (1966) and DiLorenzo (1988) also examine the predatory nature of political entrepreneurship by focusing on the rent-seeking nature of political entrepreneurs. Furthermore, Folsom (1987) identifies political entrepreneurs as businessmen who seek profits through rent seeking, whereas commercial entrepreneurs are businessmen who seek profits through market activity.

Alternatively, Schneider et al. (1995) view political entrepreneurs as agents of change that are alert to opportunities to frame debates and set voting agendas in order to pass legislation and policies. This line of reasoning builds off of the work of Dahl (1961) and Riker (1962, 1986) on how politicians and interest groups can successfully set political agendas, and emphasizes the potential societal benefits of political entrepreneurship. Political entrepreneurship, in this view, may lead to positive as well as negative social change. For instance, while a lobbyist might utilize the political system to acquire rents at the expense of taxpayers, a politician could also utilize their political power to end a costly war or halt a regulation that would hurt consumers.

The entrepreneurship theories of Kirzner and Schumpeter can also aid our understanding of (1) social entrepreneurship and (2) ideological entrepreneurship. Interestingly, the entrepreneurship theories of Schumpeter and Kirzner have not been widely cited in discussions of social entrepreneurship (Shockley and Frank 2011). Swedberg (2006, 2009), however, has argued that Schumpeter's conception of entrepreneurship can inspire and be extended to discuss social entrepreneurs creatively bringing about social change through their activities. Likewise, Shockley and Frank (2011) have argued that utilizing Schumpeter's and Kirzner's theories can advance the field of social entrepreneurship. Additionally, although there is only a nascent literature on ideological entrepreneurship,[16] Storr (2008–2009) relies on both Schumpeter's and Kirzner's theories of entrepreneurship to discuss ideological entrepreneurship.

Social Entrepreneurship

Multiple definitions of social entrepreneurship appear in the literature, but they all agree that a focus on social aims is essential for characterizing an activity as social entrepreneurship (Emerson and Twersky 1996; Light 2006; Mair et al. 2006; Martin and Osberg 2007; Wei-Skillern et al. 2007; Elkington and Hartigan 2008; Zahra et al. 2009; Dacin et al. 2010). Beyond this common motivation, social entrepreneurs are often described as constantly innovating and attempting to find new ways to serve their mission. Dees (1998: 2) has, for instance, described social entrepreneurs as "one species in the genus entrepreneur" and as "entrepreneurs with a social mission." Alvord et al. (2004) have, similarly, described social entrepreneurship as "a catalyst for social transformation." Also, Mort et al. (2003) have stressed that social entrepreneurs establish new "social enterprises" and promote innovation in existing ones. The goal of social entrepreneurship, they explain, is to solve social problems. Social entrepreneurs attempt to bring about social change by developing "innovative solutions" and mobilizing "the ideas, capacities, resources, and social arrangements required for sustainable social transformation" (Alvord et al. 2004: 262).

Although "social entrepreneurship" has become a popular concept in recent years, the range of activities that the concept describes predated the popularity of the term. Think of the individuals, motivated by religious or secular social aims, who established enterprises to give food, shelter, and clothing to the very poor. Think of the social activists and community organizers who have championed some particular social cause and are trying, through their protests, advocacy, and lobbying, to convince others of their views and win others to their cause. Think of the abolitionists who pushed for an end to slavery in the United States and the West Indies. Think of the civil rights and women's rights leaders. Think of the individuals who started almshouses and orphanages that operate independently of the government, or charitable organizations like the Red Cross, or social enterprises like the Grameen Bank. What these individuals and activities have in common is that their chief aim is social transformation, not monetary profits or political power. Though they may sometimes engage in business and political action, they have (short-term or long-term) social change agendas.

Austin et al. (2006) highlight several key aspects of social entrepreneurship and then, using Sahlman's (1996) people, contexts, deal, and opportunity (PCDO) components, point out some important similarities and differences between social and commercial entrepreneurship. Following Dees (1998), Dees and Economy (2001), Thompson (2002), and others, Austin et al. (2006: 2) define social entrepreneurship as an "innovative, social value creating activity that can occur within or across the nonprofit, business, or government sectors." They also note, however, that actual enterprises fall on a continuum between purely social and purely commercial enterprises. And, that "even at the extremes...there are still elements of both. That is, charitable activity must still reflect economic realities, while economic activity must still generate social value" (ibid.: 3).

It is also worth noting that commercial and social entrepreneurship can complement one another. Often, commercial entrepreneurs engage in social-purpose ventures. Similarly, social entrepreneurs often engage in commercial ventures. As Salamon et al. (2003) estimated, 53 percent of the revenue that nonprofit organizations receive comes from fees while 35 percent comes from government sources, and only 12 percent comes from private philanthropy. The successful businessman who uses the wealth he has accumulated to finance purely social endeavors is also a familiar figure. Social entrepreneurs may need capital accumulation from their commercial enterprises to finance their social ventures. Likewise, commercial entrepreneurs often believe that engaging in social ventures is necessary if they are to be socially responsible corporate citizens. Additionally, market activity itself can often serve as an essential tool in carrying out the social entrepreneur's mission. For instance, "Operation Fresh Start" in Madison, WI, provides contractor services to area residents and businesses but hires and trains "at risk" youths who need vocational skills and the strong mentoring that crew managers provide.[17] Social and commercial entrepreneurship are often intertwined.

Still, it might be meaningful to draw some distinctions. First, social entrepreneurship, Austin et al. (2006) contend, occurs when there is a social need that is not being met by commercial entrepreneurs. Additionally, social entrepreneurs pick up where commercial entrepreneurs fall short.[18] Social and commercial entrepreneurs create social value in different ways. Commercial entrepreneurs do so in the form of "new and valuable goods, services, and jobs, and can have transformative social impacts," but social transformation is a side effect of their activity, not their chief aim, which is necessarily "creating profitable operations resulting in private gain" (ibid.: 3). Social entrepreneurs, on the other hand, aim at "creating social value for the public good" (ibid.: 3). Compared to commercial entrepreneurs, social entrepreneurs also have difficulty attracting human and financial capital and measuring social impact (ibid.). These two challenges are, of course, related. Because it is difficult to show success, it is also difficult to attract resources and, so, to compensate staff.[19]

Although there have been considerable advances in the field of social entrepreneurship, the literature would arguably benefit from greater engagement with Schumpeter's and Kirzner's theories of entrepreneurship. Shockley and Frank (2011), for instance, argue that a theory of social entrepreneurship built on Schumpeterian and Kirznerian conceptions of the entrepreneur would have five theoretical components. Specifically, they argue, a Kirzner- and Schumpeter-inspired theory of social entrepreneurship would (1) focus on creativity and entrepreneurial discovery; (2) distinguish the entrepreneur qua entrepreneur from other agent types, including the investor, the leader, or the manager; (3) treat social entrepreneurship as a universal behavior (that can occur across sectors and across contexts); (4) treat social entrepreneurship as bringing about meaningful social change; and (5) focus on the process of social entrepreneurship and not just individual social entrepreneurs. Shockley and Frank (ibid.) argue that these components are currently underemphasized and that focusing on these components would improve theory-building within the entrepreneurship studies literature.

An alert and creative social entrepreneur can drive meaningful social change. An example might help to clarify this key role of social entrepreneurs in transforming societies. The Grameen Bank is a microfinance organization in Bangladesh that was founded by Muhammad Yunus, an economics professor, in 1983 to provide funding to the poor and spur economic development. While teaching at the Chittagong University in the 1970s, when Bangladesh was suffering from a severe famine, Yunus became interested in the economics of the rural poor. He started visiting the poor who lived near the university in order to learn more about their situation; these visits alerted him to an opportunity to help. Yunus (1999: 116) recalled that, "I was dismayed to see how the indigent in Jobra suffered because they could not come up with small amounts of working capital. Frequently they needed less than a dollar a person but could only get that money on extremely unfair terms. In most cases, people were required to sell their goods to moneylenders at prices fixed by the latter. This daily tragedy moved me to action. With the help of my graduate students, I made a list of those who needed small amounts of money. We came up with 42 people. The total amount they needed was $27."

Yunus personally lent the money to community members. Yet, when he approached the university bank about setting up a small lending program, he was told that the poor had no collateral and were therefore not creditworthy (ibid.). Yunus, however, kept lending small amounts on his own and found that his borrowers consistently paid back their loans. Still, every bank he talked to refused to start a lending program, so he started one on his own. Yunus (ibid.: 116) said, "Because I could not change the banks, I decided to create a separate bank for the impoverished. After a great deal of work and negotiation with the government, the Grameen Bank ('village bank' in Bengali) was established in 1983."

Instead of requiring collateral, the Grameen Bank requires that borrowers form small groups that share the risk of default. The small groups not only self-enforce good behavior but also provide mutual assistance and support for their members. As Yunus (ibid.: 117) describes:

> Typically a new group submits loan proposals from two members, each requiring between $25 and $100. After these two borrowers successfully repay their first five weekly installments, the next two group members become eligible to apply for their own loans. Once they make five repayments, the final member of the group may apply. After 50 installments have been repaid, a borrower pays her interest, which is slightly above the commercial rate. The borrower is now eligible to apply for a larger loan...Loan payments are made in weekly meetings consisting of six to eight groups, held in the villages where the members live. Grameen staff attend these meetings and often visit individual borrowers' homes to see how the business—whether it be raising goats or growing vegetables or hawking utensils—is faring.

The literature on the bank has found that it has opened access to credit to the rural poor and spurred social change (Bornstein 1996).[20] For instance, Dowla (2006) and Larance (2001) examine how the bank has created and fostered social capital accumulation, trust, and cooperation. Hassan (2002) and Schreiner (2003) argue that despite the expense and limits of microfinancing, the Grameen Bank appears to be broadly cost-effective as well as effective on the individual lending level. Furthermore, much of the literature examines the positive benefits of the bank on women, who have received a bulk of the loans, including reduced violence and increased use of contraception (Goetz and Gupta 1996; Schuler et al. 1995, 1996; Todd 1996).

Yunus was alert to an opportunity to provide credit to the rural poor and to drive social change. In 2006, Yunus received the Nobel Peace Prize for developing the bank and sparking a trend in microfinance. Since the establishment of the bank, other group-lending and microfinancing organizations have opened throughout the developing world. While microfinance is not the silver bullet to alleviate poverty, it has dramatically impacted the lives of the poor.

Ideological Entrepreneurship

The term "ideological entrepreneurship" is meant to describe entrepreneurship aimed at ideological change, a necessary precursor to any institutional change.[21] According to North, people only change their ideology when it no longer seems to be an adequate explanation of the world. As North (1981: 49) writes, "individuals alter their ideological perspectives when their experiences are inconsistent with their ideology." When an individual can no longer make sense of the world using his existing mental models and ideologies, he constructs or adopts new ones. But, as North (ibid.: 49) notes, it will take more than one anomaly (a single unexplainable event) to get an individual to change his ideological perspective; "inconsistencies between experience and ideologies must accumulate before individuals alter their ideology." A series of events that run counter to expectations can, thus, lead people to update their mental models and change their ideology.

Although the occurrence of repeated unexplainable events might explain the "why" of ideological shifts, it does not explain the "how" of ideological change, nor does it yield any insight into which of the array of possible ideologies will come to dominate. For North, the "intellectual entrepreneurs of ideology" or "ideological entrepreneurs" can and often do play a key role in ideological change. Repeated inconsistencies between a person's experiences and her ideology will give the ideological entrepreneur an opportunity to convince her to adopt a new theory of how the world works. As North (1981: 65) writes, when the existing ideology proves inadequate, "opportunities will be offered to the ideological entrepreneur . . . to construct a counter ideology." Also, "an ideological entrepreneur who learns of an incoherence or a disturbing implication of the ideology could utilize this in order to help reinterpret that ideology" (Denzau and North 1994: 26). Any would-be institutional entrepreneur must also be an ideological entrepreneur.

Nothing, of course, guarantees that an ideological entrepreneur will be able to convince others to accept her new ideology. North has stressed the importance of an ideology's ethical implications in determining whether or not it will be adopted. As North (1981: 49) writes, "ideology is inextricably interwoven with moral and ethical judgments about the fairness of the world the individual perceives." For instance, an ideology that points to the negative socioeconomic consequences of greed will typically be paired with an ethos that prohibits greed. Any successful counter-ideology must align with an individual's sense of right and wrong. Moreover, it must "not only provide a convincing image of the link between the specific injustices perceived by various groups and the larger system which the intellectual entrepreneurs desire altered, but," as North (ibid.: 54) writes, it must "also offer a Utopia free of these injustices and provide a guide to action."

North does offer a few historical examples of successful ideological entrepreneurs that might prove telling. Speaking of change in the ancient world, North (ibid.: 121) writes, "the agents of change too were not all kings, emperors, or their agents; they included such persons as Rabbi Akiba ben Joseph and his pupil Rabbi Meier... Jesus of Nazareth; Saul of Tarsus... and Mohammed." Rabbi Akiba ben Joseph and Rabbi Meier are noted rabbinical scholars who did much to formalize Judaism. Jesus of Nazareth and Mohammed inspired two of the world's most popular religions. And, Saul of Tarsus, the Apostle St. Paul, played a key role in spreading Christianity throughout the ancient world.

Although ideological entrepreneurs play such a key role in bringing about ideological and, thereby, institutional change, North does not devote a lot of attention to them. Beyond these examples and a few others, like Karl Marx and his popularizers and the US Founding Fathers, North leaves us guessing as to who would count as ideological entrepreneurs and who would not. He never, in fact, presents a systematic treatment of his ideological entrepreneur. We are left to guess if, and how far, his ideological entrepreneur can break a society out of path dependency. We are left to guess if his entrepreneur is a creative agent who brings about the creative destruction of ideologies, an individual who is alert to opportunities for ideological change and works to exploit those opportunities, or someone who automatically responds to relative price changes.

Storr (2008–2009), however, has attempted to develop North's ideological entrepreneur along Kirznerian and Schumpeterian lines. A Kirznerian ideological entrepreneur would be alert to opportunities to advance an existing ideology that people in a particular place want but do not yet know about (i.e., to engage in ideological arbitrage). He would also be alert to opportunities to resolve minor ambiguities or to fill in small gaps in existing ideologies. Additionally, he would be alert to opportunities to "sell" a new ideology that better explains the world than existing ideologies. Similarly, a Schumpeterian ideological entrepreneur would be a bold innovator who creates new conceptions of how the world works or combines and presents existing models of how the world works in new ways (i.e., to promote ideological development). She would work to capture the ideological marketplace, competing fiercely against other ideological entrepreneurs as well as against the weight of existing public opinion and conventions.

Although a Kirznerian and Schumpeterian ideological entrepreneur, as described above, is broadly consistent with North's "intellectual entrepreneur of ideology," there are several important advantages that our discussion of the ideological entrepreneur has over North's conception (Storr 2008–2009). First, it assigns to the ideological entrepreneur a specific and primary role in ideological and, so, institutional change. In his writings on institutional change, North has focused much less on the role of the ideological entrepreneur in changing people's beliefs and much more on the "spontaneous" changes (or lack of changes) in individuals' perceptions and beliefs that occur as a result of changes in relative prices. Second, North's ideological entrepreneur is ultimately *homo economicus* responding "automatically" and "mechanically" to opportunities to gain ideological profits. Neither Kirzner nor Schumpeter describes an entrepreneur that can be accounted for so easily within a maximizing neoclassical framework. Kirzner's entrepreneur makes genuine discoveries and Schumpeter's entrepreneur is creative and disruptive.

An ideological entrepreneur, developed along Schumpeterian and Kirznerian lines, would be able to alter a society's path. An example from Storr's research on democratization in the Bahamas will help to clarify this key role of ideological entrepreneurs in breaking societies out of path dependence (see Storr 2002, 2008–2009 as well as Martin and Storr 2007, 2009).

Until the middle of the twentieth century, a small group of white businessmen known as the Bay Street Boys exerted almost total economic, political, and social control over the majority black population in the British colony of the Bahamas. They owned almost all of the major businesses in the colony. They held an overwhelming majority of the seats in the local Assembly. Several of the businesses they owned and establishments they frequented (e.g., the Savoy movie theater and the restaurants in the British Colonial Hotel) were segregated on the basis of race. Additionally, they decided if and when Bahamian blacks would be allowed to celebrate their major cultural festivals, like Junkanoo, in the city center. The institutional matrix that existed in the Bahamas at the time was one that relegated Bahamian blacks to second-class status in the colony. There was a set of institutions that taught Bahamian blacks that, with a few exceptions and even then within limits, success for them was unlikely; this was the set of institutions that most (white and black) Bahamians believed at the time to be legitimate.

In the latter half of the twentieth century, a group of educated black ideological entrepreneurs took over the Progressive Liberal Party (PLP), then a fledgling political party, and began to challenge the legitimacy of the institutional matrix that existed in the colony (Martin and Storr 2009). Alert to an opportunity to bring about an ideological shift amongst blacks in the colony, the rhetoric of the ideological entrepreneurs in the leadership of the PLP grew more radical and divisive and their activities grew more confrontational. They were initially met with some suspicion and a great deal of resistance. Within a few years, however, they had convinced a majority of blacks in the Bahamas that the Bay Street Boys' control over the country was neither legitimate nor sustainable. Their ideological victory over the Bay Street Boys was crystallized on Black Tuesday. Protesting an attempt by the Bay Street Boys to gerrymander the constituency boundaries, and ignoring a very public warning that the government would not countenance any protest, Cecil Wallace Whitfield, others in the PLP leadership, and thousands of their supporters marched to and surrounded the Assembly on April 27, 1965. Inside the Assembly, PLP leader Lynden Pindling railed against the government's attempts to silence debate, walked to the Speaker's desk, grabbed the ceremonial mace that is meant to symbolize the power of the people represented in the Parliament, and threw the mace out of the window to the waiting crowd below. Less than two years after Black Tuesday, the PLP was the government of the Bahamas. The defeat of the Bay Street Boys at the polls also meant a defeat of colonialism, a defeat of segregation, and a defeat of artificial barriers to Bahamian black achievement. The ideological sea change that had to occur for the Bay Street Boys to lose political control over the colony cannot be overstated. The ideological entrepreneurs who controlled the PLP inspired a radical change in the thinking of Bahamian blacks. As a result of their efforts, Bahamian blacks altered their perceptions of their choice sets and began to see their current circumstances and their future prospects much differently than they had previously.

Examples of these kinds of radical ideological and institutional shifts are not rare. Of course, they do not always lead to what we would consider positive outcomes. Hitler, for instance, was an ideological entrepreneur who was alert to the growing feeling of alienation among his countrymen and convinced many of them that, in order for their nation to reach its destiny, they needed to launch a world war and begin systematically killing millions of Jews. Additionally, ideological entrepreneurs need not impact as broad a swath of people, as in the examples above, in order to be successful. The opportunities for ideological change that they recognize could be opportunities for small and subtle shifts in a particular subset of the beliefs among members of very small groups. For example, the community member who convinces her neighbors that there would be more, rather than less, social cohesion in their neighborhood if they would welcome, rather than resist, the immigrant family moving in down the street is engaging in ideological entrepreneurship. Moreover, the ideological entrepreneur can be motivated by personal as well as social gain. For instance, the businessman who convinces his clients that adopting his particular brand implies that they are sophisticated and discerning is, among other things, an ideological entrepreneur.

Alert ideological entrepreneurs can inspire a process of creative ideological destruction. They can alter paths, transforming virtuous circles into vicious circles and vice versa. Again, they can initiate positive and negative social change.

The Differences between Different Kinds of Entrepreneurs Are Not Really Differences in Kind

It is our contention that the distinctions that are typically drawn among different kinds of entrepreneurs (i.e., commercial, social, and ideological) are generally overblown. In fact, such distinctions may not hold theoretically or empirically. At the very least, the boundaries that separate different types of entrepreneurs are very porous.

To be sure, the different designations of entrepreneurs are meant to capture what are supposedly very important differences (1) in the nature of the opportunities noticed and the type of social change the entrepreneur intends and actually brings about, as well as (2) in the nature of the rules that govern their behavior in the different spheres where they operate and the knowledge surrogates, feedback, or learning mechanisms that they can rely on to guide their behavior. But, the differences within-type along these margins are arguably greater than we might suppose and the differences across-type along these margins are arguably much less clear than we might imagine.

Differences in the Nature of the Opportunities Noticed Are Exaggerated

There are supposedly sharp differences in the nature of opportunities that different kinds of entrepreneurs notice, the type of social change they intend to bring about, and what they actually do bring about through their efforts. Commercial entrepreneurs recognize, cultivate, and exploit profit opportunities and bring about economic change. Social entrepreneurs recognize and exploit opportunities for social transformation. Ideological entrepreneurs recognize opportunities to change people's attitudes and, thereby, promote institutional change.

It is, however, difficult to imagine opportunities for commercial gain, social transformation, or ideological shifts that do not have at least minimal effects in other areas. It is also easy to imagine opportunities where the commercial, social, and ideological elements are not only difficult to disentangle but also inextricably linked. Consider, for instance, the entrepreneur who establishes a commercial enterprise with a social purpose, or the entrepreneur who establishes a social enterprise but earns a profit. Examples of these kinds of enterprises are quite common.[22] Imagine also the entrepreneur who recognizes that a new product or service may be profitable precisely because of how its introduction will change the way people think and interact with each other. Simultaneous and interconnected ideological, social, and commercial revolutions arguably accompanied the introduction of the steam engine, the automobile, the vacuum cleaner, the telephone, the washing machine, the airplane, the personal computer, and the smartphone.[23]

Differences in the Feedback Mechanisms Are Not So Stark

Commercial entrepreneurs, it is argued, can rely on profit and loss to guide their operations. Markets do a good job of ensuring that members in a society get what they want, because the only way entrepreneurs can earn profits is by satisfying customers' wants. These profits not only encourage them to continue doing what they had been doing but also encourage other entrepreneurs (and would be entrepreneurs) to follow suit. When entrepreneurs fail to satisfy customers' wants, they earn losses. Losses discourage entrepreneurs from doing what they had been doing before and suggest to other entrepreneurs that they need to find different or better ways of serving customers if they hope to succeed. The feedback from profits and losses is a powerful mechanism for making sure that entrepreneurs meet consumer demands and, therefore, behave in socially useful ways.

According to this view, noncommercial entrepreneurs do not have access to such a well-functioning feedback mechanism. Earning monetary profits and incurring monetary losses cannot be viewed as meaningful signals for nonmarket entrepreneurs. Indeed, if a social need could be met profitably, there would be no need for social entrepreneurs; commercial entrepreneurs would work to fill that need. Similarly, if changing people's minds in a particular way was profitable, then a commercial entrepreneur would likely notice the profit opportunity and work to bring about the ideological shift.

There is, however, a feedback mechanism, akin to profit and loss, that can guide social as well as ideological entrepreneurs.[24] In order to be successful, social entrepreneurs must enjoy high status and a good reputation (Chamlee-Wright 2004; Chamlee-Wright and Myers 2008; Boettke and Coyne 2009a). Stated another way, successful social entrepreneurs tend to depend on community support. Social entrepreneurs who behave in socially undesirable ways eventually lose the support of their stakeholders and constituents. If they are not pursuing socially desirable ends, or if they are not achieving their stated goals, community members will stop volunteering, attending meetings, and contributing resources.

Similarly, it is certainly the case that between two ideologies that offer different but equally plausible theories of the world, the ideology that most people prefer to hold (for various reasons) will win a greater share of the ideological market. The ideological entrepreneur peddling an ideology that better satisfies the tastes of his customers will be more successful than a competing entrepreneur that is selling a less desirable ideology. Winning converts in the ideological marketplace is akin to winning customers in the commercial sphere. An ideological entrepreneur without any converts must alter either his message or the ideology he is peddling, or bear the cost of not having any converts.

Admittedly, the signals entrepreneurs must rely on in nonmarket settings are fraught with ambiguity. Chamlee-Wright (2010: 45), for instance, concedes that "relative to the robust signaling that market entrepreneurs enjoy, the signals [that nonmarket entrepreneurs must rely on] are admittedly blunt." She also warns against "elevating the non-price social learning above market social learning" (ibid.: 176). Similarly, Skarbek and Green (2011: 74) argue that "feedback in non-priced environments, even under competitive pressure, is not nearly as tight as the analogous feedback in market environments. The knowledge transmission mechanism of market pricing is unmatched in any other social method of achieving coordination among anonymous individuals dispersed over time and space."[25] While Chamlee-Wright as well as Skarbek and Green will concede that entrepreneurs in non-priced environments are not akin to ships without a map, they, nonetheless, suggest that if prices act like a global positioning system for commercial entrepreneurs, then social entrepreneurs are at best navigating by the stars.[26]

It is unclear if the social enterprises that can attract donors or if the ideological entrepreneurs who can win converts are actually satisfying "consumer" needs, much less behaving in socially beneficial ways. Indeed, evaluating success in social-purpose enterprises is complicated by the fact that the beneficiaries of the enterprise are often different people from the supporters of the enterprise. In a restaurant, for instance, the people paying for the food are generally the same people who are consuming the food. The restaurant that does not satisfy its customers, that does not deliver value for price in a cost-effective way, will incur losses and will not remain in business. The soup kitchen that does not deliver satisfying meals in cost-effective ways to its beneficiaries, on the other hand, might still attract donations and volunteers.

Additionally, although reality and experience place some limits on which ideologies individuals might believe in, it is arguably more likely that they will adopt an erroneous theory about how the world works than they would about which drinks will satisfy their thirst. Consequently, it is possible for ideological entrepreneurs to successfully peddle erroneous ideologies even when better explanations of how the world works are available.[27]

That said, the signals nonmarket entrepreneurs rely on, however, are not as ambiguous as we sometimes pretend them to be. The need to attract donations and volunteers does, in fact, play a disciplining role by signaling that somebody wants and is willing to commit resources (either time or money) to support the social enterprise.[28] Similarly, the need for the beliefs the ideological entrepreneur is peddling to be not just plausible but more plausible than existing beliefs does place limits on which views will come to dominate. Social and ideological entrepreneurship can be constrained by reputational mechanisms and accountability through close monitoring (Chamlee-Wright 2004; Boettke and Coyne 2009a). While such entrepreneurial activities may still result in negative social change, it is likely that donors and converts will eventually stop supporting socially undesirable activities.

Moreover, profits and losses are not the unambiguous guides to entrepreneurial action that they appear to be.[29] It is rarely entirely obvious if earning a loss in a particular period is a signal that the entrepreneur should change course or redouble her efforts. Likewise, earning a profit in a particular period is not necessarily a sign that the entrepreneur is on the correct path. Additionally, profit opportunities are not readily identifiable phenomena. Opportunity recognition requires interpretation, and deciding on the right course to pursue requires judgment. There has actually been a great deal of literature on the role of interpretation and judgment in opportunity recognition and exploitation. Prior knowledge and life experiences shape the opportunities entrepreneurs will discover (Arentz et al. 2013; Shane 2000; Yu 2001). Corbett (2007) utilizes the psychological contributions of Jung to show that learning asymmetries shape one's ability to recognize and interpret opportunities. Additionally, numerous cognitive abilities have been found to shape opportunity discovery, including asymmetries in creativity (Hills and Shrader 1998), risk (Mullins and Forlani 2005), and intention (Shepherd and Krueger 2002). Shane et al. (2003) identify numerous aspects of motivation that determine who has the ability to discover opportunities and develop and execute ideas, such as the need for achievement, locus of control, vision, desire for independence, drive, passion, goal setting, and self-efficacy. And,

Kaish and Gilad (1991) found that entrepreneurs tend to be information gatherers and opportunistic learners. Prices and, so, profit and loss signals can also be distorted by regulations as well as fiscal and monetary policy.

Moreover, confidence that commercial entrepreneurs will actually work to satisfy consumers' desires does not only emanate from the robustness of the signals of profit and loss vis-à-vis the signals that entrepreneurs receive to guide their action in nonmarket contexts, but also from a belief in the ability of actors to correctly assess how well a product (even an expensive and complex product) satisfies their desires. If, instead, it was believed that individuals could be systematically and repeatedly tricked into buying products that did not in fact fully satisfy their actual desires, then any faith in markets to deliver products that best satisfy our actual demands would be eroded. After all, it is not the "best" products in some objective sense that win in the market but the products that consumers believe to "best" satisfy their desires. It is not, for instance, the theory that best accounts for how the world works that wins out in the ideological marketplace but the plausible explanation of how the world works that best satisfies consumers' tastes. Similarly, it is not the social enterprises that best meet the community's needs that win out in the social marketplace but the social enterprises that are best able to attract donors and volunteers. Likewise, investors, consumers, and entrepreneurs in commercial settings, as is the case in noncommercial settings, could be mistaken about what is in their own, or society's, interests.

There Are Advantages to Collapsing the Distinct Kinds of Entrepreneurship

There are several advantages in resisting the temptation to stress the apparent differences between the supposedly different kinds of entrepreneurship. First, analyzing the work of an entrepreneur qua entrepreneur allows for examining the full set of activities that entrepreneurs engage in and their outcomes. For instance, in order to fully understand the impact Bill Gates has made on society, one must look at his commercial, philanthropic, and political endeavors. Second, a more general framework of entrepreneurship takes into account and appreciates the complex and integrated nature of social activity. When a community leader petitions the government to advance her social cause, she may be advancing a social goal as well as bringing about institutional change. Rather than trying to disentangle her activities, a broader framework can encompass the various sectors and activities she uses to pursue her goal. Third, this framework allows for a broader appreciation and understanding of the importance of entrepreneurship in all aspects of social life as well as in the social change process. Such entrepreneurs spur activity and change in mundane times (such as introducing a more efficient production process, providing a new good or service, or organizing a neighborhood association) as well as in times of crisis (such as organizing a protest against a social injustice, supplying extra gasoline or lumber after a disaster, or calling up neighbors to organize a clothing drive for a family in need).

In Summary, Entrepreneurs Can Drive Social Change

Social change, in the broadest sense, refers to any endogenous change in the social order. It could refer to dramatic social revolutions that radically transform the social structure, like the civil rights and women's suffrage movements in the United States or the introduction of Microsoft Windows and the Apple iPhone. It could also refer to small changes in the practices and patterns of behavior of individuals within a small group, like the change in shopping patterns that occur when a neighborhood grocer opens its doors or when a local charity begins providing free Sunday breakfast to the homeless in a particular park. It could, thus, refer to social transformations as locally significant as one East German escaping over the Berlin Wall during the Cold War or as globally significant as the fall of the Berlin Wall.

Entrepreneurs are a driving force behind these instances of social change. They are alert to opportunities to change society and are bold innovators who introduce new products, services, or ideas or establish new enterprises or revitalize existing ones to bring about social change.

Entrepreneurs fulfil their role as agents of social change both in mundane times as well as in times of crisis. It, therefore, is not a stretch to imagine that entrepreneurs might play a significant role in community recovery after a disaster. Indeed, we argue that the entrepreneur is a driving force behind community recovery after a disaster. The next chapter develops a theory of how entrepreneurs spur community recovery and positive social change after disasters. Specifically, after a disaster, entrepreneurs provide necessary goods and services, restore and replace disrupted social networks, and signal that community rebound is likely and, in fact, under way. Through their efforts, entrepreneurs lower the cognitive, emotional, and material costs associated with rebuilding after a disaster, clarify the benefits disaster victims can expect to receive if they were to repair or rebuild their damaged homes, and coordinate community recovery.

Chapter 3

How Entrepreneurs Promote Post-Disaster Community Rebound

Although it is commonplace to describe the entrepreneur as fearless or daring or a maverick, Schumpeter ([1942] 1976: 127) has described life in advanced commercial society as essentially "anti-heroic." As Schumpeter (ibid.: 128) explains, "success in industry and commerce requires stamina, yet industrial and commercial activity is essentially unheroic in the knight's sense—no flourishing of swords about it, not much physical prowess, no charge to gallop the armored horse into the enemy." Additionally, while it is easy to think of starting a charity as generous or noble, it is more difficult to think of starting a charity as being akin to leading soldiers against the battlements. Admittedly, depending on the environment, espousing certain beliefs could be quite dangerous, but many ideological entrepreneurs face, at most, social sanction for their preaching. Despite our efforts to analogize entrepreneurial activities to actions on the battlefield, there are obvious differences between the entrepreneur and the soldier.

Still, it is not absurd to think of entrepreneurs as heroic figures in mundane times. Successful entrepreneurs must exhibit a form of courage (McCloskey 2010–2011).[1] Courage is acting in the face of legitimate fears. Courage is not being paralyzed by adversity. Although recognizing and exploiting an opportunity by carrying out new combinations is not the same thing as charging into battle, entrepreneurs do take risks that put their reputation and livelihood on the line. As Naughton and Cornwall (2006: 73) explain, "Courage is the *sine qua non* virtue for entrepreneurs, since to undertake a new business entails a significant amount of risk—risk that can lead to major financial losses, emotional suffering, and strained and possibly broken relationships, as well as great financial gains, emotional elation, and a community of people." Similarly, McCloskey (2010–2011: 52–53) writes that "discovery and creativity depends also on the other virtues, in particular on Courage and Hope." Courage, however, does not necessarily mean benevolence. Heroic entrepreneurs can have a wide range of motivations, including profits, prestige, and power, as well as good will or social justice. For our purposes, it is

The original version of this chapter was revised.
An erratum to this chapter can be found at DOI 10.1007/978-1-137-31489-5_10.

not an entrepreneur's motivation that matters, but that she is alert to opportunities that promote social change.

If entrepreneurs can be thought of as heroic in mundane times, they can certainly be viewed as heroic figures following a disaster when the possibility of failure is likely to be more glaring, the adversity they must overcome to succeed is likely to be more extreme, and the level of uncertainty they must confront is likely to be more pronounced. Courageous entrepreneurs can spur post-disaster community recovery.

Community Rebound after Disaster Is a Collective Action Problem

Disasters present a significant challenge for displaced individuals. The decision to return and rebuild a damaged or destroyed home is a costly one, and the benefits of rebuilding are necessarily uncertain. Whether returning and rebuilding makes sense depends on a number of factors, including whether the goods and services that disaster victims need to rebuild are available and whether others in their disrupted social networks will also return and rebuild. In such a scenario, the rational move for any displaced individual may be to wait and see what others are going to do. The costs associated with being the first mover are arguably prohibitive. Since every displaced resident faces a similar calculus, community recovery (i.e., the return and recovery of displaced residents) after a disaster is a very real challenge.

Modeling this scenario is straightforward.[2] It can be modeled as a two-player single-shot game under conditions of imperfect information. Consider a community comprised of two families (Players A and B) whose homes were destroyed by a disaster and were subsequently displaced to different locations. Or view these players as representative disaster victims or as groups of disaster victims who have been displaced to different locales. The key feature is that communication between the players is difficult, and potentially impossible, so they cannot form reasonable expectations about the other's decision-making. The likely increased levels of regime and information uncertainty after a disaster further complicate a player's ability to guess the other's likely behavior.[3]

Players A and B both must decide whether to return and rebuild their old homes or establish new lives in a different community. Both players, however, have imperfect information about the other's intentions. The cost of returning and rebuilding is c.[4] This represents any cost of returning and rebuilding above and beyond the cost associated with remaining in exile, since players would also incur some cost when continuing to stay in the place where they were exiled.

If both players decide to return, then they each earn a benefit of α.[5] The net benefit each would receive from returning is, thus, $\alpha - c$. Any player that does not return earns a payoff of β. If either player decides to return but is the sole returnee, that player earns a benefit of δ, where $\alpha > \delta$. See figure 3.1 for a representation of this game in extensive form. It is important to note that it is the perceived costs and benefits that matter; it is what people believe them to be, not what they actually are, that guides their decision-making.

If the costs associated with returning are prohibitively high or if players would not prefer to return even if they could re-create the community that they had before the disaster, so that $\beta > (\alpha - c)$, then remaining in the new locale would be the dominant strategy for both players, and neither players returning would be the socially optimal outcome. When this condition holds ($\beta > (\alpha - c)$), disasters that displace residents will result in socially desirable depopulation.

We might alternately assume that both players would want to re-create the community that they had before the disaster if they could do so, such that $(\alpha - c) > \beta$. If $\beta < (\alpha - c)$ and $\beta > (\delta - c)$, both players returning to rebuild would be the socially optimal outcome, since by assumption $(\alpha - c) > \beta > (\delta - c)$ and $2(\alpha - c) > 2\beta > (\delta - c) + \beta$. Given that both players have imperfect information about what the other player is likely to do, however, the strategy that dominates will depend on whether the expected net value of returning is greater or less than the payoff from remaining in the new locale. This poses a serious challenge for individuals in communities affected by disasters. Players do not have ready access to the decisions, circumstances, or intentions of others, because connecting with others who are also displaced is extremely costly, if not impossible. This challenge is even more severe the larger the number of families in the community that were displaced by the disaster and the more widespread the evacuation destinations. As such, most families form expectations about what other displaced residents will do based on (at best) incomplete information. This is true of the two-player game depicted in figure 3.1.

Assuming that the expected net value of remaining in the new locale is greater than the expected net value of returning under conditions of

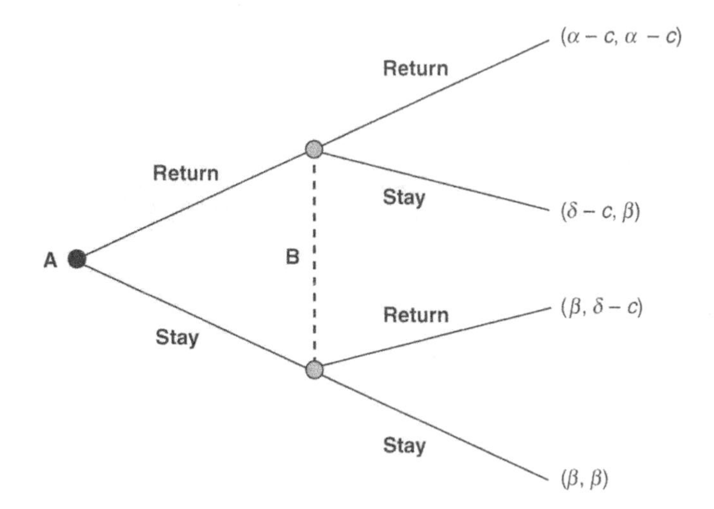

Figure 3.1 Return game with imperfect information.

persistent uncertainty, remaining in the new locale will be the dominant strategy. That is, assuming $\beta > (\alpha - c)P_r + (\delta - c)(1 - P_r)$ where P_r is the player's judgment as to the probability (i.e., the perceived probability) that the other player will return, the dominant strategy will be to not return. Although not returning is the dominant strategy, where $2(\alpha - c) > 2\beta$, it is socially suboptimal.

In this scenario, there are several things that would change the nature of the game or alter the payoffs such that community rebound is socially optimal and likely to occur (i.e., returning and rebuilding becomes the dominant strategy). For instance:

- One of the players could locate the other players, encourage them to return, and convince them to credibly commit to returning once they decide to do so (i.e., increasing P_r).
- A third party could increase the benefits associated with returning by offering to provide public, club, or private goods and services to the players once they return (i.e., adding θ, where θ is a newly offered good or service).
- A third party could lower the cost associated with returning by offering to assist by subsidizing the rebuilding process or offering a lower priced alternative to needed goods and services (i.e., adding γ, where γ is a newly offered subsidy or the savings associated with the availability of the lower-priced good or service).
- A third party could connect displaced residents with one another and encourage, coordinate, or compel the players to return (i.e., increasing P_r).

This list is clearly not exhaustive. But, taken together, the items on this list suggest that it is possible for the players themselves or a third party to change the game depicted in figure 3.1 into the game depicted in figure 3.2.

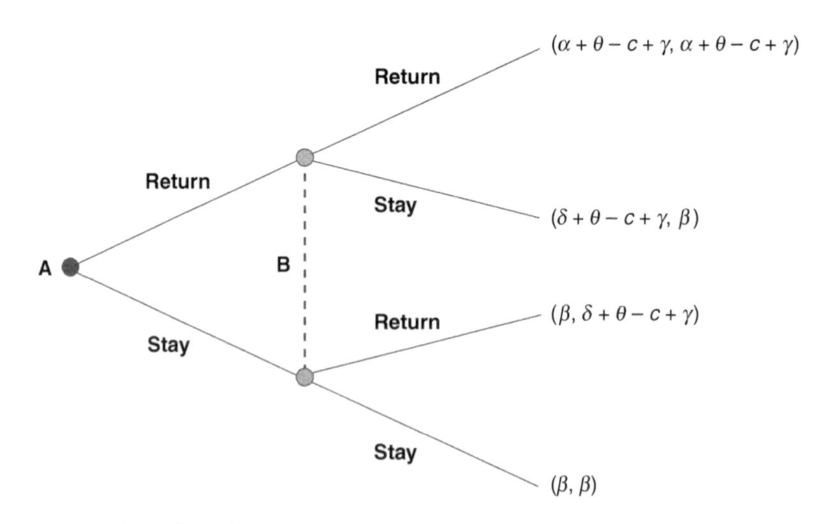

Figure 3.2 Transformed return game.

When $(\alpha+\theta-c+\gamma)\,P_r+(\delta+\theta-c+\gamma)\,(1-P_r)>\beta$ returning and rebuilding is the dominant strategy, and community rebound is the socially optimal outcome.[6]

Disaster Victims Do Expect Governments to Help Them

Disaster victims and affected communities often look to the government to assist them with returning and rebuilding after a disaster.[7] Specifically, after a disaster, governments can provide financial assistance to displaced residents and businesses, as well as restore and replace damaged or disrupted publicly provided goods and services. However, governments can also retard recovery by implementing plans or adopting rules that lower the expected benefits or raise the expected costs associated with rebuilding and returning or reduce the expected likelihood that others would return.

There is a considerable literature on the important role the government plays in post-disaster recovery, both at the federal and local levels. For instance, Pipa (2006) argues that a high-level coordinating body (namely, a federal government agency) is needed to manage the activities of different actors, including nonprofits, after a disaster. Similarly, Cigler (2009: 765) concludes that "the 9/11 and Katrina tragedies highlighted the importance of a strong FEMA [Federal Emergency Management Agency]."[8] Schneider (2008) and Thaler and Sunstein (2008) also conclude that a centralized response is necessary to deal with major disasters. According to Thaler and Sunstein (2008: 13), "As the experience with Hurricane Katrina showed, government is often required to act, for it is the only means by which the necessary resources can be mustered, organized, and deployed." Additionally, Coppola (2015: 476) notes, "the crucial point, often overlooked, is that disaster risk management is a full-time, multi-dimensional responsibility that should be exercised throughout a government and evident across a society." Likewise, both Tierney (2007) and Springer (2009, 2011) call for governments to adopt an integrated emergency management system and preparedness framework, improved communication strategies, and strengthened oversight. As Springer (2009: 1) argues, "Improving disaster response capabilities within [the United States]...requires better coordination not only within the Department of Homeland Security, but also across the federal government as well as with state and local governments and the private and non-profit sectors." Consistent with Springer's recommendations, Fakhruddin and Chivakidakarn (2014: 160), when discussing the role for government in disaster management, argue that "not only technical capacity for early warning, therefore, but also institutional capacity in policy planning and disaster management are essential if national disaster response strategies can be assessed as being effective."

The argument that a centralized authority is necessary to prepare for and respond to disasters is often tied to the notion of moral hazard. While it is in everyone's interest to prepare for and seek insurance against potential disasters, doing so is costly. Some residents will shirk, hoping that others will be better prepared and that they can reap the benefits. However, if everyone shirks, society will not be prepared. Instead, governments can take on the burden of preparedness and thus overcome the moral hazard issue. For instance, Besley (1989) argues that government-provided disaster insurance can overcome moral hazard issues associated with private insurance. Burby (2006) also finds a role for federal government to enforce mitigation and preparedness efforts at the local level.

Furthermore, much of the literature that emphasizes the importance of local efforts also calls for federal coordination and oversight. For instance, both Burby (ibid.) and King (2007) suggest that better government planning and more comprehensive local government plans would help to speed up post-disaster recovery. As Chamlee-Wright (2010: 2) explains, "particularly in the context of New Orleans, the paradigm that shaped the public policy discourse was that a large-scale government response was the obvious and only remedy to solve the problems presented by catastrophic disaster."

Following a large disaster, the government's ability to coordinate the activities of hundreds of thousands of displaced residents, build community confidence that recovery will succeed, and spur the return of residents is severely weakened because it faces a similar coordination problem to the one that displaced residents face. While it is rational for individuals to wait for utilities to be turned on, roads to be cleared, support services to be restored, infrastructure to be rebuilt, and neighbors to return, it is also likely that—because of the enormity of the disaster and the multiple potential uses for limited government resources—it makes sense for the government to wait for its affected citizens to commit to returning and rebuilding before restoring utilities, clearing roads, and so on. Admittedly, if policymakers could pick viable neighborhoods and identify and support likely first movers, they might be able to overcome this coordination problem and encourage community confidence and spur return. Or, if enough resources were available, the post-disaster coordination problem that exists between government officials and displaced residents could be made irrelevant.

Social Capital Also Plays an Important Role in Post-Disaster Recovery

Disaster victims often deploy their reserves of social capital after a disaster. Indeed, social capital can be an important resource following a disaster, allowing disaster victims to access material support as well as aiding them in making predictions about the likelihood that others will return and rebuild. There is now considerable literature on the important role social capital plays in post-disaster community recovery.

Social capital is a resource that facilitates collective action for mutual benefit. Bourdieu (1985), who offered the first systematic treatment of social capital within social science, has described it as those resources that individuals have access to as a result of their possessing durable networks of relationships. Coleman (1988), who popularized the concept among sociologists, has similarly described social capital as the collection of resources that exists within and as a result of the relationships "between actors and among actors." Although some have critiqued the concept as being too broad, and various definitions of social capital have been utilized since the concept was first introduced, there is now largely a consensus over the meaning of social capital (see, for instance, Foley and Edwards 1999; Portes 2000; Adler and Kwon 2002). Social capital describes the resources that are associated with membership in some particular network that individuals use as they pursue their goals.

The literature identifies that social capital comes in the form of social networks, norms, and shared narratives.[9] The literature also identifies three types of social capital: bonding, bridging, and linking (Woolcock 2001). Bonding social capital refers to the strong links that exist between like-minded individuals in homogenous groups. Think of networks comprised of family members, friends, or the members of a close-knit community (e.g., the congregation of a neighborhood church, the pledge classes of a sorority, etc.). Bridging social capital refers to the weak ties that exist between the members of heterogeneous groups. Think of the links between people who live together in a diverse neighborhood or between members of the same sorority that joined in different years. Linking social capital refers to the connections that exist between individuals from completely different social settings or varying positions of power within the community. Think of the links between a mentor and a mentee in an inner city mentorship program or between certain residents and the town mayor. Not surprisingly, different types of social capital are better at facilitating different types of activities.

Researchers have found that there is a positive relationship between various types of social capital and various measures of societal well-being.[10] Social capital, for instance, has a positive impact on economic growth (Knack and Keefer 1997; Woolcock 1998; Torsvik 2000), civic engagement and the enhancement of government performance (Putnam 1995; Fukuyama 2001; Poulsen and Svendsen 2005; Carden et al. 2009; Chong et al. 2010), crime reduction (Putnam 1993; Dilulio 1996), educational attainment (Coleman 1988), and health (d'Hombres et al. 2010). Likewise, Granovetter (1973, 1983, 1995) emphasizes the strength of weak ties over strong ties when describing how individuals pursue their plans and improve their social and economic status. Putnam (1995), conversely, worries that a decline in social capital has led to civic disengagement and widespread distrust in democratic institutions.

Not surprisingly, social capital has also been found to aid in disaster preparedness, response, and recovery (Aldrich 2011a, 2011b, 2012; Bolin and Stanford 1998; Chamlee-Wright 2010; Chamlee-Wright and Storr 2009b, 2009c, 2010a, 2011b; Paton 2007; Pelling 1998; Pelling and High 2005; Shaw and Goda 2004). There is a strong, positive, well-documented relationship between the existence of social capital and community recovery.[11] Studying the post-Kobe earthquake experience as well as the 2001 earthquake in Gujarat, India, for instance, Nakagawa and Shaw (2004) find that communities with high levels of social capital were more efficient in rescue, relief, and recovery. They also find that communities with high levels of social capital tended to be happier with the post-disaster development planning process.

Several of these studies have focused on the key role that bonding social capital plays in facilitating community return. Chamlee-Wright and Storr (2009c), for instance, discuss the role that bonding social capital played in the swift return of the Vietnamese-American community in New Orleans East after Katrina. In less than two years after Katrina, the closely connected networks comprised of the congregants of the Mary Queen of Vietnam (MQVN) Catholic Church in New Orleans East had facilitated the return of that neighborhood by giving community members access to a variety of resources, which they used to overcome the social coordination and political difficulties associated with rebuilding their distinct ethnic–religious–language community after Katrina.[12]

Similarly, Hurlbert et al. (2000, 2001) find that individuals embedded in networks that contain higher proportions of family members had greater access to informal support and were more likely to activate these core networks for support preceding and after a disaster, like in the case of Hurricane Andrew, than individuals embedded in networks that did not have high proportions of kin. Likewise, Aldrich (2011a, 2011b, 2012) describes the important role bonding social capital in combination with ties to trans-local institutions (i.e., linking social capital) played in the post-tsunami recovery of southern India.

Focusing on two cases, the 2003 blackout in the eastern part of Canada and the United States that affected 50 million people and the 2000 E. coli crisis in Walkerton, Ontario, Murphy (2007) argues that networks of both strong and weak ties help communities rebound from hazards and disasters. Murphy (ibid.), for instance, finds that over a third of the people surveyed reported that they provided assistance to their neighbors, family members, and even (albeit in rare cases) strangers during the blackout. And, an even larger percentage reported that they believed their neighbors would have provided assistance to them had they needed it. Social networks, Murphy (ibid.) also concludes, played an important role after the Walkerton crisis. Residents of this small, close-knit community, where an extensive stock of social capital existed before the crisis, reported that they relied on both strong ties (i.e., close family and friends) and weak social capital (i.e., distant family and acquaintances) during the crisis.

While Hurlbert et al. (2000, 2001), Aldrich (2011a, 2011b, 2012), and Murphy (2007) acknowledge that networks comprised of weak ties (particularly links to extra-community institutions) can be important, they nonetheless stress the importance of bonding social capital in bringing about post-disaster recovery.[13]

There are several reasons why we should expect that individuals living in heterogeneous, loosely connected communities would find it harder than individuals living in homogenous, tightly connected communities to overcome the collective action problem associated with post-disaster recovery.[14] First, it is harder to develop accurate expectations around the decisions of others when they are less well known (i.e., in communities comprised of individuals with only weak ties between them). Arguably, this problem is magnified the greater the cultural, ethnic and social differences as well as the more novel the circumstance. For instance, predicting whether a neighbor will mow his lawn on the weekend is easier than predicting whether he will return and rebuild after his home is destroyed. Second, discovering the intentions of others is more costly the weaker the association (i.e., finding out what your spouse versus what your neighbor will do in a particular situation).[15] The difference in costs is potentially greater after a disaster where it is possible that an individual may not even know how to contact weaker ties.

Community-based organizations, however, have been found to play a key role in helping loosely connected communities leverage the extra-community connections and other resources of their members. Community-based organizations can help coordinate the recovery of even heterogeneous, loosely connected communities by ensuring that community members are kept aware of developments in the devastated area and of the likelihood their neighbors will return. In addition to working to funnel resources for their communities, community-based organizations perform two key functions to help heterogeneous, loosely connected communities overcome the collective action problem that complicates post-disaster recovery. First, by collecting and sharing information about community members' plans and the various challenges facing the community as a whole, community-based organizations make it more likely that community members will form accurate expectations about each other's decisions. Second, by keeping in regular contact with community members and organizing community meetings and other activities, they make it easier for community members to coordinate around community rebound. Such community-based organizations utilize bridging and linking capital, or weak ties, to communicate with displaced residents, acquire resources from outside the community, and spur recovery.

Bolin and Stanford (1998), for instance, explore how community-based organizations aided in recovery after the 1994 Northridge, California, earthquake and found that community organizations provided direct assistance to low- and moderate-income earthquake victims and worked with government agencies to funnel resources for the repair, rehabilitation, and development of homes in these communities. Similarly, Shaw and Goda (2004) find that Kobe residents created community organizations that performed a similar function after the 1995 Kobe earthquake in Japan (i.e., working with local agencies to influence the recovery planning process). Likewise, Murphy (2007) found that a majority of respondents also reported having volunteered during the Walkerton crisis with preexisting and newly established community service organizations to provide more vulnerable members of their community with bottled water and other resources. And, Storr and Haeffele-Balch (2012) establish that, in the Broadmoor neighborhood of New Orleans, residents and business owners, through the organization of the

Broadmoor Improvement Association, communicated electronically via text messages and emails with displaced neighbors to determine who was returning and how they could help one another to get the neighborhood back on its feet after Hurricane Katrina.[16]

Researchers have also explored how network characteristics affect disaster response and how disasters change networks. Dynes (2006), for instance, has explored how social capital changes after disasters. Networks, he finds, become restructured after a disaster. Networks that developed around accomplishing goals related to work or family become reoriented toward helping individuals recover from the disaster. Additionally, obligations become reordered following a disaster. Pre-disaster obligations matter less than the post-disaster obligations to render assistance that develop. Social networks also become a critical source of information in the fluid, confusing circumstances following a disaster. Furthermore, following a disaster, established community organizations take on new roles and new community organizations are created that aid in recovery. See, also, Adler (2010) for a similar study of how social capital is changed by natural disasters.

Studies on the role of social capital in disaster preparedness, response, and recovery have tended to focus on social capital as social networks. But social capital in the form of collective narratives can also play an important role in bringing about post-disaster recovery. Collective narratives give us insight into the schemas (Sewell 1992) or first-order constructs (Schutz 1973) that community members use to interpret their circumstances and decide on a course of action. As Gerteis (2002: 609) concludes, "collective narratives are important because they are the sites where schemas take concrete empirical form." After a disaster, social capital in the form of shared narratives helps communities to (1) make sense of their circumstances, (2) assess their capabilities and prospects for recovery, and (3) decide on, and work toward, a course of action. If community members believe themselves to be powerless, their circumstances to be grim, and their prospects to be hopeless, then community recovery is likely to be retarded, regardless of the assistance they are provided. Similarly, if community members believe themselves to be resilient, their circumstances to be manageable, and their prospects to be hopeful, then community recovery is likely to progress even if they do not have access to all the resources they might want.

Collective narratives that encourage self-reliance (even if external assistance is forthcoming) and celebrate the community's history of overcoming challenges, then, would tend to facilitate community recovery. This is not to suggest that social capital in the form of socially beneficial collective narratives is the only productive asset that community members need to possess. Rebuilding a house requires know-how, building materials, tools, and labor. Interpretive schemas alone, however favorable, are unlikely to provide any of these necessary resources. But, absent an interpretative frame that sees rebuilding as possible, rebuilding efforts will be hampered. Interpretive schema that lead communities to work toward recovery as soon as possible after a disaster rather than waiting for external aid, and that allow them to link the current disaster with a difficulty the community has withstood in its past, enhances the community's capacity to recover. As Chamlee-Wright and Storr (2010c) show, collective narratives that stress the community's capacity for overcoming challenges was an important factor in the swift recovery of the MQVN community, where community members, in addition to utilizing the available community resources, linked their Katrina experience with their migration experience from North Vietnam.[17] Similarly, as Chamlee-Wright and Storr (2011b) argue, the shared narratives held by parishioners in St. Bernard Parish before Katrina and which emerged after the storm (especially because theirs is a tight-knit community, comprised of hardworking, blue collar workers) aided in St. Bernard's recovery.[18]

A Community's Capacity to Rebound Is Related to Its Capacity for Self-Governance

The characteristics that make for a successful city neighborhood in mundane times are also likely to make for resilient city neighborhoods in difficult times. Stated another way, those neighborhoods that engage in effective self-governance in normal times are also likely to have the capacity to effectively respond to and recover from disasters.

There has been considerable literature on how communities overcome complex challenges, like managing common-pool resources or dealing with crime, providing public goods or overcoming problems of neighborhood blight, or responding to the challenges associated with post-disaster recovery. Elinor Ostrom, for instance, has written extensively on common-pool resource (CPR) scenarios.[19] CPR scenarios are situations in which a tragedy of the commons could lead to overuse of a resource. E. Ostrom, however, describes how local communities devise and enforce rules to monitor the use and maintenance of a scarce resource and, therefore, overcome the tragedy of the commons. E. Ostrom's (1965) initial study focuses on the development of public enterprises as well as associations to minimize saltwater intrusion into the groundwater basin near Los Angeles, California. In addition to exploring several diverse natural resource management scenarios, E. Ostrom and others have also explored metropolitan organizations (Advisory Commission on Intergovernmental Relations 1987, 1988, 1992; V. Ostrom et al. 1961; V. Ostrom et al. 1988), the provision of public goods (V. Ostrom and E. Ostrom [1977] 1999), and privatization in developed and developing countries (Oakerson et al. 1990).

Although there are no blueprints when it comes to solving CPR scenarios, there are similarities among successful self-governing CPR institutions. As E. Ostrom (1990) explains, robust CPR institutions tend to have clearly defined boundaries, appropriation rules that are consistent with local conditions, arrangements for making collective decisions, effective monitoring, graduated sanctions for rule violators, effective mechanisms for conflict resolution, and the ability of community members to organize without outside interference to solve their commons problems.

The greater degree of "community" a group exhibits, the more likely that group is to develop robust institutional regimes that solve collective action problems (E. Ostrom 1992). Stated another way, the greater degree of "community" a group exhibits (before a challenge), the more likely that group is to solve complex problems like post-disaster recovery. Where "appropriators share generalized norms of reciprocity and trust" and "the group appropriating…is relatively small and stable," the greater "the likelihood of CPR appropriators adopting a series of incremental changes in operational rules to improve joint welfare" (E. Ostrom 1990: 211).[20] While shared norms as well as small and stable social networks are not by themselves sufficient for the emergence of robust CPR institutions, they are positively correlated with the other group characteristics that E. Ostrom suggests might be more important. For instance, appropriators are more likely to share a "common judgment" about the problem and face lower enforcement costs when networks are stable and norms are shared.[21]

Vincent and Elinor Ostrom have also written about how communities work to solve complex challenges within polycentric orders and, specifically, why complex community challenges are more likely to be solved within polycentric orders than in monocentric orders. A polycentric order is one in which individuals are able to organize into multiple governing authorities at different scales; that is, there are multiple, formally independent, decision-making entities (E. Ostrom 1999). These governing authorities can be governmental or nongovernmental bodies, and their jurisdictions can, and often do, overlap. Although each of the multiple governing authorities enjoys a degree of autonomy in decision-making, the governing authorities within a polycentric order are constrained by overarching rules. This setting is in contrast to monocentric orders (or what V. Ostrom et al. (1961) refer to as "gargantua"), which consist of a single decision-making body.

Community challenges are more likely to be solved within polycentric orders because authorities within these orders are more adaptable and flexible than authorities within monocentric orders. Individuals within a large society have different preferences and different knowledge. Although V. Ostrom et al. (1961: 838) explain that "no a priori judgment can be made about the adequacy of a polycentric system of governance as against the single jurisdiction," there are advantages to polycentrism in dynamic scenarios. As we discuss in chapter 8, there are reasons to believe that polycentric orders will be more effective at promoting post-disaster community recovery than monocentric orders.

Although there are certainly differences between post-disaster recovery and managing a common-pool resource, both demonstrate a community's capacity for self-governance. The community with the capacity to develop effective mechanisms for dealing with common-pool resources is likely to have the capacity to develop effective responses to post-disaster challenges.[22] Additionally, the degree of "generalized norms of reciprocity and trust" in a community is likely to be related to the community's ability to both manage its common-pool resources as well as solve the collective action problems associated with disaster recovery. Moreover, local governing authorities (like community organizations and associations) operating within polycentric orders are more likely to solve the challenges associated with common-pool resources and disaster recovery because of their access to local knowledge and their greater flexibility than more centralized governing authorities. Arguably, community groups can outperform gargantua in both the management of common-pool resources and post-disaster community recovery.

Jacobs (1961) has described city neighborhoods as the loci of self-government. As such, successful city neighborhoods are ones where neighbors effectively respond to neighborhood-wide problems when they arise, and unsuccessful city neighborhoods are ones where neighbors fail to adopt solutions to their collective problems. Viewed this way, neighborhoods are not judged based on the quality of their public areas, homes, or schools. Instead, they are to be judged based on how effectively they react to neighborhood-wide concerns. As Jacobs (ibid.) explains, it is possible for rich city neighborhoods to be unsuccessful and poorer communities to be successful at dealing with neighborhood-wide challenges as they arise. Similarly, neither better housing nor better schools necessarily lead to better behavior among inhabitants. Good houses and schools deteriorate in unsuccessful neighborhoods while poor houses and schools are improved in successful neighborhoods (through the collective efforts of their inhabitants). As Jacobs (ibid.: 114) argues, "city neighborhoods [are] mundane organs of self-government. Our failures with city neighborhoods are, ultimately, failures in localized self-government. And our successes are successes in localized self-government. I am using self-government in its broadest sense, meaning both the informal and formal self-management of society."[23]

Jacobs (ibid.) highlights several characteristics of good communities, including having an array of public characters with meaningful links to key and often quite powerful individuals beyond their neighborhoods and throughout their districts.[24] Stated another way, effective communities are populated by community leaders whose social networks give them access to important bridging social capital. Jacobs (ibid.) refers to these leaders as hop-skip people and their links as hop-skip links.[25] As Jacobs (ibid.: 135) explains, "a city district requires a small quota of...people who know unlikely people." And, "these are working relationships among people, usually leaders, who enlarge their local public life beyond the neighborhoods of streets and specific organizations or institutions and form relationships with people whose roots and backgrounds are in entirely different constituencies, so to speak" (ibid.: 135). These links facilitate information transfer about available resources and resource needs, as well as facilitate the provision and acquisition of available and needed resources.

Social networks in city neighborhoods, however, have to exhibit a certain degree of stability if such neighborhoods are to be successful. If neighborhoods are to be cohesive, Jacobs (ibid.: 138) explains, "if self-government in the place is to work, underlying any float of population must be a continuity of people who have forged neighborhood networks. These networks are a city's irreplaceable social capital. Whenever the capital is lost, from whatever cause, the income from it disappears, never to return until and unless new capital is slowly and chancily accumulated." Social networks in successful city neighborhoods can accommodate new residents and even some turnover, but those "increments and displacements have to be gradual" (ibid.: 138). Stated another way, there needs to be a stable reservoir of bonding social capital if city neighborhoods are to be effective.

As the Ostroms and Jacobs explained, a community's capacity for self-governance depends on (1) the social coordination capacity of community organizations and associations, (2) the ability of community members to effectively access both bonding and bridging social capital, (3) the ability of community members to leverage their shared histories and perspectives, and (4) the stability of social networks within the community. The greater a community's capacity for self-governance, the better able it is to deal with complex challenges such as crime, the provision of public goods, or problems of neighborhood blight. Arguably, these pre-disaster systems of self-governance can, and do, aid in post-disaster community recovery.

There have been several studies that examine the characteristics that make communities resilient in the face of disasters.[26] For instance, Cutter et al. (2010) suggest that there are numerous factors that affect disaster resilience, including higher levels of college education, home ownership, and formal preparedness and mitigation plans, as well as lower levels of minorities, the elderly, and disabled persons. Similarly, Burton (2014) tested numerous characteristics using data from Hurricane Katrina and finds that higher levels of per capita income, home ownership, and housing density, as well as more cultural, religious, and social advocacy centers and professional service occupations all correlate with higher community resilience. He also finds that lower levels of minorities were correlated with higher community resilience. These findings highlight the importance of economic development and broader prosperity—including accumulations of wealth, education, and social capital, as well as the existence of quality institutions and infrastructure—for resilience. These findings also point to the importance of a community's pre-disaster capacity for self-governance in explaining its post-disaster capacity for recovery.

Entrepreneurs Help Community Members Overcome the Collective Action Problem That Plagues Community Rebound

Entrepreneurs also play a critical role in helping communities to rebound after a disaster. Specifically, entrepreneurs perform at least three key roles: (1) they provide needed goods and services; (2) they help to restore or replace disrupted social networks; and (3) they signal that community rebound is likely and, in fact, under way.

Entrepreneurs Provide Needed Goods and Services

Perhaps the most obvious and least controversial role entrepreneurs play after a disaster is the provision of goods and services. After all, entrepreneurs provide goods and services during mundane times and, thus, it is reasonable to believe that they will perform this function in other situations as well.

Repairing, rebuilding, or replacing damaged or destroyed homes after a disaster requires, among other things, building materials and technical skills. Building materials have to be secured, either by being purchased or donated. Contractors, carpenters, plumbers, and electricians need to be enlisted. Additionally, disaster victims have to replace damaged clothing, furniture, appliances, and other household goods. Moreover, people trying to rebuild after a disaster need goods and services at levels not typically required in regular times. If local stocks of building materials were able to satisfy predisaster demand, for instance, they are bound to be inadequate, even if they were not also damaged or destroyed during the disaster, once rebuilding efforts are under way. Similarly, if the pool of skilled laborers in a particular locale was large enough to satisfy local demands before the disaster, it is bound to be too small, even if these laborers were not displaced by the disaster, once returning residents begin employing them to rebuild their damaged or destroyed homes.

Entrepreneurs recognize this increased demand for goods and services and move to satisfy it. To return to the language introduced above, entrepreneurs increase the benefits associated with returning (i.e., adding θ, where θ is the offered good or service) by providing goods and services to disaster victims once they return and begin rebuilding their damaged or destroyed homes. Entrepreneurs also lower the costs associated with returning (i.e., offering γ) by offering a lower priced alternative to the needed goods and services and creating new enterprises or reorienting existing enterprises to assist displaced residents in the rebuilding process.

There is, in fact, a growing literature on entrepreneurship in hostile environments, including places affected by terrorism, civil war, and natural disaster (Bullough et al. 2013; Galbraith and Stiles 2006). Although one might assume that the added uncertainty of these scenarios eliminates the possibility for entrepreneurship, studies illustrate that entrepreneurs are active in these environments (Bullough et al. 2013). Entrepreneurs, for example, are enticed by higher prices and seize this opportunity to make a profit, or are motivated by the heightened suffering experienced by disaster victims and attempt to relieve that suffering. Other studies suggest that the psychological traits of entrepreneurs make them better suited to cope with adverse environments compared to others (Baron and Markman 2000; Branzei and Abdelnour 2010; Joseph and Linley 2008; Markman et al. 2005; Tedeschi and Calhoun 2004).

There are several reasons why entrepreneurs might be well positioned to provide necessary goods and services after a disaster. For instance, entrepreneurs may have access to local knowledge, including knowledge of local conditions (such as the specific needs of particular affected residents) as well as of available resources that survived the disaster and can perhaps be utilized or repurposed for disaster recovery. Entrepreneurs utilize these skills and experiences to provide goods and services after a disaster, spurring community rebound.

Entrepreneurs Help Community Members Restore or Replace Disrupted Social Networks

In addition to providing needed goods and services, entrepreneurs help community members restore, repurpose, and even replace social networks that were disrupted by the disaster. Remember that disasters can disrupt as well as destroy social networks and that social networks play a critical role in helping individuals recover from disasters. As Aldrich (2012: 2) explains, "higher levels of social capital—more than such factors as greater economic resources, assistance from the government or outside agencies, and low levels of damage—facilitate recovery and help survivors coordinate for more effective reconstruction." And, "deeper reservoirs of social capital serve as informal insurance and mutual assistance for survivors, help them overcome collective action constraints, and increase the likelihood that they will stay and work to rebuild (as opposed to moving elsewhere)" (ibid.: 2).

Entrepreneurs can discover and create opportunities for displaced residents to connect and share information with one another. In some cases they act as a critical connection among the displaced portions of a social network that was tightly connected before the disaster (i.e., the key link between two subgroups of once closed social networks). Entrepreneurs can also connect disaster aid demanders and disaster aid suppliers, including linking disaster victims with the scores of donors and volunteers who typically respond to disasters. Additionally, entrepreneurs can create and supply social spaces where disaster victims can meet and form, or re-form, links with one another. This is critical after a disaster where regular (i.e., pre-disaster) social spaces (i.e., homes, churches/temples, restaurants, social clubs, etc.) are likely to be scarce.

Entrepreneurs can, thus, help to coordinate return by working to restore disrupted social networks and facilitating the creation of new ones. To return to the language introduced above, entrepreneurs reduce the costs associated with returning and rebuilding, and also increase the ability of disaster victims to coordinate their return by connecting displaced residents with one another and other sources of support. (i.e., adding γ or increasing P_r).

Entrepreneurs Signal That Community Rebound Is Likely and, in Fact, Under Way

In addition to providing needed goods and services and helping community members restore or replace disrupted social networks, entrepreneurs can also act as "focal points," or "points of orientation," for residents as they formulate their plans to return and rebuild (Chamlee-Wright 2010). Recall that the decision to return is intimately tied to the decisions of others, and that post-disaster community rebound is a collective action problem. As Schelling (cited in Gosselin 2005) remarked shortly after Hurricane Katrina, "it essentially is a problem of coordinating expectations...If we all expect each other to come back, we will. If we don't, we won't." But Schelling did not think that there was a private solution to disasters of the scale and scope of Katrina. According to Schelling (ibid.), "There is no market solution to New Orleans...achieving this coordination in the circumstances of New Orleans...seems impossible...There are classes of problems that free markets simply do not deal with well...If ever there was an example, the rebuilding of New Orleans is it."

There is, however, a Schelling solution to the problem that Schelling identified. As Schelling (1960) explains, a focal point acts as a single equilibrium around which players' expectations can converge in a game where there are multiple Nash equilibria. A focal point, then, is a signal that the players of the game recognize and believe other players will also recognize. As such, it is a signal that the players of the game can coordinate around. Think of traffic lights. Think also of shared beliefs that might act as focal points. As Schelling (1957: 21) explains, focal points provide "some clue for coordinating behavior." In game situations, focal points tend to have "some kind of prominence or conspicuousness" (ibid.: 22).

Entrepreneurs may act as focal points after disasters by signaling that specific communities will return and rebuild. They can do this in a number of ways. In sequential games, like the collective action problem following a disaster, a first mover might be an adequate signal that community return is under way. Entrepreneurs can strategically act as first movers by accepting the risks associated with returning before the recovery process is really under way, and also by laying the groundwork for return by reopening or creating the enterprises that will service, supply, and hire displaced residents after recovery is actually under way.[27] Entrepreneurs can also work to coordinate recovery by recognizing opportunities to facilitate communication among displaced residents and organizing recovery efforts, as well as by creating new enterprises or reorienting existing ones to eliminate the barriers to community recovery. Finally, simply by resuming activities, entrepreneurs signal a return to normalcy and so allow disaster victims to coordinate around return. Stores opening for business, social clubs meeting to discuss this or that, and churches holding services all signal that life in a post-disaster context is returning to how it was before the disaster.

To return to the language introduced above, entrepreneurs can coordinate return by connecting displaced residents with one another, encouraging displaced residents to return and rebuild, organizing recovery efforts, or signaling that recovery is under way (i.e., increasing P_r).

In Summary, Entrepreneurs Are Heroic Figures after Disasters

Entrepreneurs have proven to be heroic figures after disasters. The post-disaster environment is a difficult one for them to negotiate. Still, through their efforts, they help communities overcome the collective action problem that characterizes post-disaster community recovery. Specifically, entrepreneurs provide necessary goods and services, help community members restore or replace disrupted social networks, and signal that community rebound is likely and, in fact, under way. While it is possible to view almost everything that everyone does in the post-disaster context as entrepreneurial, we highlight these three roles as being particularly important for recovery. Together, entrepreneurs performing these three roles spur disaster victims to coordinate on returning by increasing the net benefits associated with return and reducing the uncertainty regarding how others will respond to the disaster.

It is important to note that, in addition to promoting community recovery after disasters, several of the entrepreneurs that we discuss in this book were actually heroes in the classic sense. Consider, for instance, the actions of Casey Kasim and Father Vien during, and in the immediate aftermath, of Hurricane Katrina.[28] Casey Kasim did not evacuate before Hurricane Katrina, choosing instead to stay in New Orleans with his wife to guard his gas station and convenience store in the Lower Ninth Ward. After the storm surge, Kasim used his boat to rescue over 30 people who were trapped in their homes, and he and his wife provided food and supplies for survivors until they could safely leave the area. Similarly, Father Vien stayed in New Orleans East with approximately 500 of his parishioners, taking up shelter in the MQVN Catholic Church. Many residents went home after the storm to survey the damage, leaving them vulnerable when the floodwaters rushed into their neighborhood. Father Vien and a group of volunteers used boats to rescue the stranded residents, and they all took shelter on the second floor of the church until it was safe to leave. Both Kasim and Father Vien acted quickly to help their neighbors, risking their own lives to help those in need.

Chapters 5, 6, and 7 will focus on the entrepreneurial efforts that helped bring about community recovery after Hurricane Katrina and Hurricane Sandy. Hurricane Katrina struck the Gulf Coast of the United States on August 29, 2005, causing a storm surge and severe flooding in New Orleans, Louisiana, after 53 levees failed. All in all, Katrina caused over 1,800 deaths, resulted in over $100 billion in damage, and displaced over 400,000 Gulf Coast residents. Hurricane Sandy hit the east coast of the United States on October 29, 2012, causing over 160 deaths and over $60 billion in damage. The next chapter provides details on both storms and covers our methodological approach to examining entrepreneurship after disasters.

Chapter 4

How Entrepreneurship Promotes Community Recovery: The Cases of Hurricanes Katrina and Sandy

There has been an increase in the reported number of natural disasters throughout the world in recent years. According to the Centre for Research on the Epidemiology of Disaster (CRED), the number of natural disasters worldwide has steadily increased, with 24 in 1950; 65 in 1975; 296 in 1990; and 528 in 2000.[1] CRED identifies 11 different types of natural disasters, including droughts, earthquakes, epidemics, extreme temperatures, floods, insect infestations, landslides, mass movements, storms, volcanic activities, and wildfires.[2] Since 1950, there have been 12,813 natural disasters worldwide. Of these natural disasters, 62 percent were either floods (34 percent) or storms (28 percent).

The increasing number of natural disasters has led to significant property damage. Total damages from natural disasters amounted to $1.96 billion in 1950; $2.03 billion in 1975; and $8.93 billion in 2000.[3] The number of people affected by natural disasters worldwide has also been increasing, with 10.8 million people affected by natural disasters in 1950, 33 million in 1975, and 172 million in 2000.[4] Although more people have been affected by natural disasters, there has not been any increase in deaths as a result of natural disasters.[5]

Stromberg (2007), also using CRED data, advances several hypotheses for the increase in the number of natural disasters. First, he cites the findings of the Intergovernmental Panel on Climate Change, which attributes more extreme weather to climate change and predicts more extreme weather in the future. Second, Stromberg (ibid.) points to possible changes in reporting. For example, closed, undemocratic societies underreport disasters, so if countries are becoming more open, then the number of reported natural disasters could increase. Third, population growth may also contribute to more shocks being considered natural disasters, based on the way in which CRED defines natural disasters. As populations increase, it becomes more likely that a shock will either affect at least 100 people or lead to ten fatalities (which is the criteria used by CRED to determine when a shock should be labeled a disaster).

Unsurprisingly, natural disasters have hit some parts of the world harder than others. Again, considering the period from 1950 to 2000, many of the reported natural disasters hit parts of Asia, with 11 reported in 1950 (46 percent of the total number of disasters), 26 in 1975 (40 percent), and 195 in 2000 (37 percent). Some of this can be attributed to the lower elevation in many parts of Asia, which leads to increased risk of flooding. Kahn's (2005) research corroborates this finding. He finds that geography does matter when considering the number of natural disasters that a country or region experiences, and indicates that Asia has a disproportionate number of natural disasters. Kahn (2005) also finds that rich and poor nations have the same quantity and severity of natural disasters but that richer countries have fewer fatalities. There are a few explanations for this lower number of fatalities. One explanation is that richer countries tend to have safer buildings that can withstand some levels of flooding (including flooding from storm surges), earthquake tremors, and other types of natural disasters. Another explanation is that richer countries have more advanced warning systems to alert citizens of potential dangers.

CRED indicates that the number of disasters in the United States has also been increasing since 1950. In 1950 and 1975 there were four disasters each, and in 2000 there were 31 disasters. Data on natural disasters within the United States is also reported by FEMA, and although CRED figures and FEMA figures follow the same trend, FEMA reports many more severe weather events.[6] According to FEMA data, in 2013 there were 61 major disaster declarations (compared to 28 disasters reported by CRED), five emergency declarations, and 28 fire management assistance declarations (for a total of 94 severe weather events).[7] There are at least two reasons for this difference. First, CRED and FEMA use different definitions to classify disasters. Second, as the literature on FEMA disaster assistance indicates, FEMA data is more likely to be influenced by political factors (Garrett and Sobel 2003; Sobel and Leeson 2006; Leeson and Sobel 2008; Shughart 2011).[8]

In our research, we focus on two disasters in the United States, Hurricane Katrina in 2005 and Hurricane Sandy in 2012. We focus our study on natural disasters in the United States for several reasons. One reason is access to data. We were able to access both quantitative and qualitative data from Katrina and Sandy, including hundreds of interviews conducted with individuals affected by the disasters. In addition to the interview transcripts, we have access to and information about the communities in which the natural disasters took place. In each case, the authors spent weeks on the ground, walking around the neighborhoods, talking informally to local residents, and conducting formal interviews. Following Hurricane Katrina, for instance, our research team spent considerable time in the Greater New Orleans Region during the period from March 2006 to January 2009. Similarly, we conducted fieldwork in New York in July 2013 and July 2014. Researchers asked the disaster victims a variety of questions, such as: What social networks and elements of civil society were present before the disaster? What was the role of government in helping their community recover? Had the community experienced other hardships in the past? If so, how had they responded to those events? What expectations existed about resources for response and recovery? Were there any shared narratives that unified expectations or helped contribute to recovery? And, what was the role of entrepreneurs in helping displaced residents return and rebuild after disasters? Information gathered in the field was corroborated with data found in academic books and articles, news sources, and reliable data sets to create an understanding of what the community was like before the disaster and how individuals responded following the disaster. We contend that without this detailed information from fieldwork, we would be unprepared to offer a thorough analysis of the role entrepreneurs play in post-disaster community recovery.

Also related to data availability, quantitative data for disasters in the United States and disasters in other parts of the world are difficult to compare. As we mentioned, FEMA tracks the number of declared disasters in the United States, the number of fatalities, and the extent of the property damage. Comparable data from other countries often do not exist, especially data from developing countries. In developed countries, for instance, property insurance payouts and pre-disaster property valuations are used to estimate damage; however, often no such data exist for disasters in developing countries. Further, even if property damage estimates did exist, trying to meaningfully compare the estimated property loss of a home in New York and the estimated property loss of a home in a less developed country would be problematic. Although the monetary valuation of property in New York would greatly exceed that in a developing country, an argument could be made that in terms of the effect on well-being, the home owner in the developing country is worse off following the destruction of her home compared to the home owner in New York (who often has personal savings, or friends and family who have the resources to take her in temporarily, and in doing so, mitigate the impact of the disaster on her living standards).

Admittedly, even in our collection of case studies, there is variation that makes comparison somewhat difficult. We do not pretend that, for example, the population of the Lower Ninth Ward in Orleans Parish, Louisiana, is the same as the population of Far Rockaway in Queens, New York. We do, however, contend that there are certainly commonalities in the challenges that the victims of disasters encounter, the strategies that disaster victims adopt to rebuild and recover, and the role entrepreneurs play in promoting recovery in the case studies that we examine. Although there are sure to be some commonalities in the recovery efforts after the 2013 floods in the northern Indian state of Uttarakhand and after Hurricane Katrina in the Greater New Orleans Region, trying to construct meaningful parallels between the recovery efforts is more challenging and, on many margins, inappropriate. This is the case even though the number of lives lost in the flood in Uttarakhand and the number of Katrina victims are comparable.[9] We argue that context is key to understanding a disaster as well as disaster recovery. Relevant contextual information might include income levels and standard of living, the pre-disaster systems of mutual assistance, and the role of local, state, and national governments in everyday affairs. Seen this way, it is likely that Uttarakhand, India, and New Orleans, Louisiana, have such divergent characteristics that they are likely to lead to divergent recovery strategies.

Below, we describe Hurricane Katrina and Hurricane Sandy. By studying these disasters we believe that it is possible to gain a better understanding of the challenges disaster victims encounter, the strategies they adopt, and the role entrepreneurs play in promoting community recovery. We then describe the research methods utilized in this study.

Hurricane Katrina Devastated the Greater New Orleans Region

The level of devastation wrought by Hurricane Katrina and the flooding that followed is unmatched in US history. Hurricane Katrina was the costliest hurricane to hit the United States, with a total estimate of damages reaching over $100 billion (Knabb et al. 2006). Hurricane Katrina affected Louisiana, Alabama, Mississippi, and the southern tip of Florida. New Orleans, Louisiana, suffered most of the damage as a result of levee breaches that led to flooding throughout the city. As much as 80 percent of the city was flooded, and over 134,000 housing units, or 70 percent of all occupied housing units, were damaged.[10] Tragically, over 1,800 people died as a result of Hurricane Katrina and the subsequent flooding.

New Orleans is built atop swampland and much of the city is between four and six feet below sea level. To the east of New Orleans is Lake Borgne, which opens out to the Gulf of Mexico, Lake Pontchartrain is to the north, and the Mississippi River weaves in from the northwest, traveling south of the city and into the Gulf. The Mississippi River Gulf Outlet (MRGO) levee runs down the western boundary of Lake Borgne. At the top of the levee is the Intracoastal Waterway, which then connects to the Industrial Canal, running north to south, joining Lake Pontchartrain and the Mississippi River. As Hurricane Katrina hovered over the city, the storm brought two devastating surges of water.[11] One surge, caused by the counterclockwise swirl of the storm, entered Lake Borgne and toppled the MRGO levee, sending floodwater into the city from the east. A second surge was caused by storm winds in the Gulf and northern winds across Lake Pontchartrain. The resulting water penetrated the Industrial Canal, London Avenue Canal, and 17th Street Canal, causing levee breaches, the most significant of which were along the Industrial Canal.

Communities bordering Lake Pontchartrain—such as Lakeview, Gentilly, and New Orleans East—experienced over ten feet of flooding. Breaches along the Industrial Canal closer to the Mississippi River, where a surge from Lake Borgne overcame the MRGO levee, brought some of the most severe flooding (also in excess of ten feet) to the Lower Ninth Ward. Helicopter views of the city showed only lines of rooftops, with cars, boats, and other debris floating on the water. An estimated 60,000 people, who had not evacuated, had to be rescued from rooftops. The US Coast Guard played an important role in rescue efforts, rescuing nearly 33,500 people and helping to evacuate almost 10,000 patients.[12] Many residents were evacuated to shelters out of state, including in Houston, Texas.

The scale of the disaster required response efforts unparalleled in US history. Hundreds of thousands of people suddenly were in need of basic food assistance, clean water, clothing, medical care, and temporary housing. FEMA, however, took as long as a week to arrive. As the Emergency Management Director of the Jefferson Parish, Walter Maestri, explained, "For approximately six days we sat here waiting" (quoted in Phillips 2005).

Of the total of $120.5 billion in federal funds spent following Hurricane Katrina, almost $75 billion went to emergency relief (ibid.). Additionally, total philanthropic giving was over $6.5 billion.[13] Hundreds of churches from around the country donated food, clothing, and volunteer hours. As a result of the severity of storms, the extent of flooding, and the damage to homes, as many as 273,000 people were evacuated and housed in temporary shelters before, during, and after Hurricane Katrina. According to the Red Cross, the organization provided 1,400 evacuation shelters for survivors, many outside the state of Louisiana. The organization also served over 68 million hot meals and snacks.[14] Similarly, the Southern Baptist Church committed to provide 300,000 meals a day for three months (*PBS* 2005). The larger church organizations communicated with local congregation leaders to understand what was needed on the ground. Reverend Edwards of the United Methodist Bethany Church in Gentilly, for instance, mentioned that congregations as far away as California contacted him wanting to help. Moreover, as Horwitz (2009a) notes, private businesses, especially big-box stores, played an important role in the disaster response efforts. Wal-Mart arrived in New Orleans ahead of FEMA, and, between August 29 and September 16, 2005, delivered some 2,500 truckloads of goods. Further, drivers and trucks were available to help move donations into effected areas. Later, as many as 114,000 households were housed in temporary FEMA trailers.[15]

Unfortunately, many of the pumps that were needed to drain the city were damaged or destroyed by Hurricane Katrina. As a result, in some cases, communities remained under floodwater for several weeks. The water that flooded the city was not salt water, or even gray water—it was black water, which contained dangerous bacteria and other toxins. To safely remove debris and start cleanup, residents had to wear masks and other protective gear. Following Hurricane Katrina, the Centers for Disease Control (CDC) randomly inspected homes affected by the storm and found that as many as 46 percent had visible mold.[16] Mold presents both a health risk and creates additional challenges in the cleanup process. Walls with mold had to be gutted and, in some cases, entire homes had to be leveled.

Following the disaster, many disaster victims found themselves unable to begin recovery. Some were not allowed to return to their communities because their neighborhoods were deemed unsafe by authorities. Those who could and did return were confronted with a series of challenges. As

Chamlee-Wright (2010: 1) describes, "when the first residents returned to New Orleans, the circumstances were anything but hospitable. Storm debris blocked roads, there was no access to clean water, no electrical service, none of the area businesses were open, the contents of homes and businesses had been festering for weeks, and the mold was advancing by the day." An estimated 118 million cubic yards of debris had to be removed throughout the city. If the debris had been stacked onto a football field, it would have reached a height of over 10.5 miles high (*Fox News* 2006). Home owners also had to dispose of and replace furniture, appliances, and other personal belongings that were damaged because of floodwaters.

As noted above, because of the scale of the disaster, social networks were disrupted. Some residents evacuated to nearby cities before Hurricane Katrina hit. Others who had attempted to weather the storm were taken by bus or plane to nearby cities and, in some cases, more distant states. For instance, as many as 250,000 evacuees ended up in Houston, Texas, after the storm.[17] Additionally, for those whose homes were badly damaged or destroyed, it was difficult to return to the city because they had to find alternative housing while they repaired and rebuilt their homes. Recall that over 400,000 residents were displaced after the storm (Geaghan 2011). While FEMA did provide temporary trailers to disaster victims, the application process took time and was not always easy to navigate. As a result, many communities were conspicuously empty in the months following Katrina.

As was discussed in the previous chapter, recovering from a disaster like Hurricane Katrina requires overcoming a major collective action problem. While some of the costs associated with returning and rebuilding in the wake of Katrina were arguably easy to determine, others were more difficult to estimate. For instance, disaster victims may know the value of their damaged houses and lost possessions, but they may not know how difficult it will be to obtain a building permit or how much paperwork will need to be filed to get insurance payments or assistance from FEMA. Similarly, the benefits associated with returning and rebuilding were far more ambiguous. Since part of the benefits that displaced residents would hope to attain were associated with their social ties, each individual's decision to return and rebuild was (at least partially) dependent on the decisions of their family members, friends, and neighbors. If others planned on returning, it might make sense to also plan to return. If others decided to permanently settle elsewhere, returning might not be the sensible option. Under such a scenario, the likelihood of a community successfully rebounding will depend, in part, on the ability of community members to signal to each other that recovery is likely.

Not surprisingly, in the wake of Katrina, some communities were better able to overcome this collective action problem than others. See table 4.1 for details on the pace of the post-Katrina recovery.

There has been something of a checkerboard pattern to recovery in New Orleans post-Katrina. Indeed, New Orleans recovery has been uneven and often slow. As Chamlee-Wright (2010: 2) describes, in the first few years following Katrina, "some [New Orleans] communities demonstrated robust signs

Table 4.1 Repopulation as percentage of pre-Katrina population in the Greater New Orleans region through August 2014

	July 2005 Pre-Katrina Population	August 2006 (% of Pre-Katrina Pop.)	August 2008 (% of Pre-Katrina Pop.)	August 2010 (% of Pre-Katrina Pop.)	August 2012 (% of Pre-Katrina Pop.)	August 2014 (% of Pre-Katrina Pop.)
Orleans Parish	198,232	98,141 (49.51%)	142,846 (72.06%)	159,418 (80.42%)	171,614 (86.57%)	176,265 (88.92%)
Lower Ninth Ward (ZIP code 70117)	20,191	6,150 (30.46%)	10,026 (49.66%)	12,038 (59.62%)	13,468 (66.70%)	14,622 (72.42%)
Gentilly (ZIP code 70122)	18,233	4,462 (24.47%)	9,684 (53.11%)	12,292 (67.42%)	14,225 (78.02%)	14,684 (80.54%)
Broadmoor (ZIP code 70125)	9,119	2,975 (32.62%)	6,070 (66.56%)	6,359 (69.73%)	7,003 (76.80%)	7,281 (79.84%)
Village de l'est in New Orleans East (ZIP code 70129)	4,919	1,839 (37.39%)	2,819 (57.31%)	3,166 (64.36%)	3,308 (67.25%)	3,373 (68.57%)
St. Bernard Parish	25,604	N/A	11,761 (45.93%)	14,149 (55.26%)	15,520 (60.62%)	16,092 (62.85%)
Chalmette (ZIP code 70043)	12,969	N/A	5,528 (42.62%)	7,067 (54.49%)	7,951 (61.31%)	8,117 (62.59%)

Note: Each neighborhood or city, except Chalmette, is included in a ZIP code that covers a greater geographical area. Chalmette encompasses the entirety of ZIP code 70043.

Source: The Data Center Analysis of USPS Delivery Statistics Product, "Residential addresses actively receiving mail by ZIP code and parish for the New Orleans metro area," http://www.datacenterresearch.org/.

of recovery right from the start; others appeared to be caught in a state of suspended animation, with residents and community leaders unable to gain significant momentum forward." Similarly, as Weber and Peek (2012: 2) explain, "more than five years after the storm, tens of thousands of former Gulf Coast residents remain displaced. Some are desperate to return to the region but do not have the means. Others have chosen to make their homes elsewhere. Still others found a way to return home but could not find a way to remain." Approximately one year after Hurricane Katrina (August 2006), the population of Orleans Parish was 98,141 (compared to a population of 198,232 in July 2005), or 50 percent of the population of the city prior to the storm. In August 2010, the population had increased to 159,418, or 80 percent of the estimated population in July 2005 (also see Schigoda 2011). While the French Quarter and Garden District in August 2010 had 95 percent of the number of active households that it had pre-Katrina, Central City had 78 percent, Gentilly Woods had 65 percent, and Desire had only 41 percent of the number of active households that it had pre-Katrina.

Hurricane Sandy Caused Major Damage

Hurricane Sandy hit the East Coast of the United States on October 29, 2012, and barreled down on the densely populated states of New Jersey and New York. Hurricane Sandy hit New Jersey first, making landfall just north of Atlantic City. The storm devastated the central and northern New Jersey coast with much of the damage caused directly by the storm surge. The storm surge pushed water through the New York Bay and up the Hudson River, bringing floodwaters to Jersey City. The highest storm surge measured was at Kings Point on Long Island Sound, measuring 12.65 feet above normal tide levels.[18] In the United States, 73 people were killed; there were over 180 deaths in total, including one fatality in Canada and the remaining in the Caribbean.[19] Hurricane Sandy resulted in over $60 billion in property damage—the second costliest disaster is US history (after Hurricane Katrina).

In New Jersey, prior to the storm, mandatory evacuations were issued for the barrier islands from Sandy Hook to South Cape May. However, the storm brought a level of destruction few could have imagined. As many as 72,000 buildings were damaged or destroyed, with 37,000 being primary residences. The majority of the buildings damaged were in Ocean County, in central New Jersey, which contains a long stretch of barrier islands, including the communities of Seaside Heights and Long Beach Island. Following the storm, 2.7 million people were without power in New Jersey. FEMA estimated that, in the state of New Jersey alone, there was over 8.7 million cubic yards of debris.[20]

As Hurricane Sandy moved north, it plowed into Staten Island, New York. Powerful waves and storm surge devastated the coast, and floodwaters moved through the city as the surge traveled up New York Bay and the Hudson River. In New York City, approximately 305,000 homes were destroyed or damaged, mostly by the storm surge, and over 40,000 residents were temporarily displaced.[21] The estimate for private and public property damage in New York was $19 billion. Flooding caused significant damage to infrastructure and public transportation.[22] Major transportation networks were suspended, including New York's Metropolitan Transportation Authority (the subway system), as well as Kennedy Airport and LaGuardia Airport.

Total estimated property damage as a result of Hurricane Sandy was $60 billion. FEMA provided approximately $7 billion in assistance, with $1.2 billion provided through the Individuals and Households Program (IHP), $5 billion through Small Business Administration (SBA) loans, and approximately $800 million through Public Assistance (PA).

The scale of damage resulting from Hurricane Sandy suggested that those areas most severely affected would take time to recover. Although FEMA had substantial resources available to help through the IHP Program, there were initially difficulties in getting the funds to disaster victims. Before Hurricane Sandy, FEMA policies stipulated that money could only go to home owners, not renters. With millions of renters in New York City, however, excluding this population meant that many disaster victims would not be eligible for funds. Consequently, new policies were adopted to assist renters. The Disaster Housing Assistance Program (DHAP) was created for home owners (through the US Department of Housing and Urban Development) and the Temporary Disaster Assistance Program (TDAP) for renters (through the New York City Department of Housing Preservation and Development).

Because of the level of flooding, planners and other officials tried to reevaluate the boundaries of existing flood plains. Similar to the scenario following Hurricane Katrina, existing National Flood Insurance Program maps proved to be outdated. However, these efforts took time and subsequently caused delays in rebuilding, while disaster victims waited for the assessment. In December 2012, Advisory Base Flood Elevation (ABFE) maps were released by FEMA. The new maps were released for key counties in New Jersey, including Atlantic, Hudson, Monmouth, and Ocean Counties.[23] Changes to flood maps also affected residents in New York. The ABFE maps included information about how high homes had to be elevated in certain areas. Further, new federal legislation, the Biggert-Waters Flood Insurance Reform Act of 2012, changed flood insurance rates.

The population affected by Hurricane Sandy was diverse. Indeed, images in the days and weeks following the storm presented multimillion dollar mansions in New Jersey destroyed by the storm surge and, in other places, including the Rockaway Peninsula, New York, low-income individuals stranded in large public housing complexes. Seaside Heights, a resort community in Ocean County, New Jersey, was devastated. This community of approximately 3,000 people and a median household income of just over $31,000 (according to the 2010 census) also hosts a popular boardwalk and amusement park, much of which was washed away by Hurricane Sandy. Hurricane Sandy further twisted segments of roller coasters, which, after the storm moved through, were left precariously standing as ocean waves crashed against the structures. Piles of wood from piers and the boardwalk were strewn across the beach.[24]

In Staten Island, New York, 23 people were killed, more than any other borough in New York City, and peak storm tides reached 16.5 feet. Staten Island is a borough of approximately 470,000 people, with a median household income of $72,500 per year (according to the 2010 census). The borough is home to many blue-collar workers, including policemen, firefighters, and other emergency personnel.

Most of the post-Sandy entrepreneurs that we highlight in subsequent chapters reside and work in the Rockaway Peninsula, which includes the communities of Far Rockaway and Bayswater (population of approximately 60,000). Residents of the Rockaway Peninsula experienced up to six feet of flooding. Those living in basement apartments lost all their possessions. The community was without power for two weeks, meaning that streets were completely dark at night, and, as October came to a close, individuals and families had to find ways to stay warm. Further, damage to infrastructure, including roads, buses, and the subways, made an already isolated community even more isolated. Gas shortages made transportation problems worse.

The recovery in New Jersey and New York has been mixed. Population data at the level of borough in New York City masks the effects of Hurricane Sandy, because of the size of the five boroughs and transient nature of the city.[25] Table 4.2 provides some figures on how many households were affected in New York and New Jersey, and the status of repairs one year later.

Different community members recovered from Hurricane Sandy at different paces, with low-income residents taking longer to recover. As a report by *Enterprise Partners* (2013: 7) found, "across the tri-state area, a greater share of low-income households are contending with unrepaired housing damage compared to those with greater means. In New York City, for example, 76% of non-low-income households are living in housing where all damage has been repaired, compared to only 52% of low-income households."

Furthermore, different communities recovered at different paces. In Seaside Heights, New Jersey, for instance, Coolidge Avenue was still mostly deserted one year after Hurricane Sandy. As a resident explained, where there had previously been 35 homes, 23 remained vacant in October 2013 (Schapiro

Table 4.2 Percentage of households with damage after Hurricane Sandy and repair status one year later

	Percentage of Households with Reported Damage %	Percentage of Households with No Repairs as of Oct. 2013 %	Percentage of Households with Repairs Partially Complete as of Oct. 2013 %	Percentage of Households with Repairs Completed as of Oct. 2013 %
New York	14	10	26	64
New York City	15	10	25	65
New Jersey	27	11	33	57

Source: Enterprise Partners (2013: 4), https://s3.amazonaws.com/KSPProd/ERC_Upload/0083708.pdf.

2013). Similarly, parts of the boroughs of Queens and Staten Island were devastated after the storm. Two years later, in October 2014, Breezy Point in the Rockaway Peninsula was still not fully recovered. Of the 348 homes destroyed (218 by storm surge and 130 by fire), only 20 households had been able to move into new homes (Trager 2014).

Programs have been developed to help disaster victims rebuild or, in some cases, relocate to areas with less risks of flooding. For instance, a New York state buyout program was available to residents in Staten Island as a way to incentivize home owners to relocate. In August 2014, 248 properties had been purchased in Staten Island as part of the program (Sherry 2014). And, in October 2014, 50 homes had been demolished (McEnery 2014). The New York City Build It Back program, which provides funding for home owners and renters, had little impact until a program overhaul in 2014.[26] By October 2014, the program had made some progress, "Nearly half (almost 6,500) of Build it Back applicants have now been made an offer by the program—compared to only 451 earlier this year; Over 4,100 have now accepted an offer—compared to 0 earlier this year; Over 1,500 have now started design—compared to 0 earlier this year; And 762 have moved into construction and another 1,090 have received reimbursement checks—both compared to 0 earlier this year."[27] Similarly, there has been no shortage of criticism over New Jersey's assistance program for home owners (the Homeowner Reconstruction, Rehabilitation, Elevation, and Mitigation), for delays, confusing application processes, and poor management of funds. In New York, recovery has also been slow in places like Staten Island, where residents make note of excessive red tape.

In Order to Examine Community Recovery after Disasters, We Adopted a Primarily Qualitative Approach

As we will show in the following chapters, entrepreneurs played a critical role in helping the communities affected by Hurricane Katrina and Hurricane Sandy rebound. We should note that the analysis presented here is part of a larger study of the political, economic, and social factors affecting disaster preparedness, response, and recovery. The broader study focuses on how entrepreneurs promote community rebound and how people are making use of (or failing to make use of) resources embedded within their social networks to contribute toward effective community recovery. The study began as the *Crisis and Response in the Wake of Hurricane Katrina* project, a five-year study of the rebuilding and recovery efforts following Hurricane Katrina, funded by the Mercatus Center at George Mason University and conducted by Mercatus Center scholars and researchers. As part of the project, hundreds of interviews were conducted in Louisiana, Mississippi, and Texas. Since then, Mercatus Center scholars and researchers have continued to study post-disaster recovery, including examining community rebound after Hurricane Sandy. Additionally, researchers have looked at community recovery after the 2011 tornado outbreak in Tuscaloosa, Alabama, and Joplin, Missouri.

They compared the two disasters to better understand the effects of different recovery strategies.

Along with the quantitative and theoretical analysis that informs the project, qualitative methods were also deployed. The choice of qualitative methods and, in particular, semi-structured interviews is particularly helpful in addressing the issue of how people carve out strategies for rebuilding, since many of the strategies people have adopted would have been impossible to anticipate in advance, rendering a survey methodology inadvisable.[28] The semi-structured interview format also creates a window through which the researcher can see more clearly which social norms, cultural tools, and other socially embedded resources are being deployed as residents and other stakeholders carve out strategies for individual recovery and community rebound.

Chamlee-Wright (2010: 27) likens understanding the post-disaster scenario and recovery process to putting together a complex puzzle. Imagine, she asks, that the puzzle begins as 24 regular puzzle pieces. It seems reasonable, she suggests, to assume that a researcher could easily assemble the 24 pieces into a recognizable image. She asks us to then imagine that gradually the puzzle becomes more complicated and the 24 pieces become 1,000 pieces and the two-dimensional figure becomes three-dimensional. The puzzle then grows and grows so that it covers the space of a modern city. Next, she asks us to imagine that the puzzle starts to change. Some of this change is the result of exogenous forces—a crane moves more pieces into the scene. Others, however, are endogenous. The pieces themselves reflect on their position and move somewhere else; they form strategies, have expectations, and learn.

As Chamlee-Wright (ibid.) explains, an aerial view of the puzzle may help us track changes over time. We can see which parts are growing the fastest. Determining these broad generalizations about the puzzle might require engaging in econometric methods. If we want to know why the pieces are choosing to move in the direction and the speed that they appear to be moving, however, such methods are less helpful. As Chamlee-Wright (ibid.: 29) notes, "qualitative research is like stepping inside our puzzle, so that we can understand what constitutes the environment in which our puzzle pieces (or people) are operating. We can observe individual behavior and the interactions between these living and interpreting beings."[29] Qualitative methods, including in-depth semi-structured interviews, give the researcher access to local knowledge, and by studying and coding the data, the researcher can uncover patterns regarding why and how certain decisions were made and certain actions were pursued. Indeed, during our interviews, we asked a variety of questions designed to understand the particular details of the person's disaster experience, how they navigated that experience, and the meaning they assigned to the experience.

The research team conducted over 300 interviews in Orleans and St. Bernard Parishes. An additional 53 interviews were conducted in Houston, Texas. Houston was selected because there was a large group of individuals from New Orleans who had been evacuated to Houston (sometimes forcibly). Months and even years after Hurricane Katrina, many former New Orleans residents remained. We wanted to better understand why these former New Orleans residents had elected to stay in Houston or whether they had plans to return to New Orleans. Within the Greater New Orleans area, we interviewed individuals in a number of different communities. Some of those communities were (in no particular order) Algiers, Chalmette, Kenner, Central City, Gentilly, St. Bernard Parish, Broadmoor, New Orleans East, the Ninth Ward, and the Lower Ninth Ward. The interviews were conducted between March 2006 (approximately seven months following Hurricane Katrina) and January 2009.

The research team deployed a mixture of random sampling and purposive sampling to select subjects. Early interview subjects were asked to connect us to people in their social networks. We make no claims to the sample being representative. In cases where the purposive sampling method was used, we opted for this approach, given how few people had returned to the area. For example, purposive sampling was used in St. Bernard Parish, due to the fact that, in the period of interest, less than two out of every five residents had returned. Furthermore, the majority of those residents who had returned were living in trailers next to damaged houses and spending any free time they had working to repair them. The effectiveness of traditional methods of sampling (i.e., random) was inevitably limited in this context. Moreover, since we were concerned with exploring social networks and the narratives community members embraced as they worked toward recovery, it made sense to interview multiple people connected through the same network nodes. However, to avoid speaking to only a narrow band of the social spectrum, the research team complemented its purposive strategy with random sampling within particular neighborhood blocks.

The research team attempted to speak to a variety of stakeholders, including residents, government officials and workers, business owners, and nonprofit leaders. And, there were unique interview guides for (1) residents, (2) faith-based groups and nonprofit organizations, and (3) business owners. As noted above, all interview guides followed the general pattern of asking about the community pre-Hurricane Katrina, what happened during the storm, how and why the individual decided to come back, what the community has been like since Katrina, and if the individual could identify and describe the role of any leaders in the community.

Interviews typically lasted between 30 minutes and an hour and a half. Once a round of interviews was completed, the audio files were transcribed. The transcripts were then coded for themes and patterns, including those discussed here regarding the role of entrepreneurs in post-disaster recovery.

The same strategies and similar interview guides were used following Hurricane Sandy. Interviews conducted following Hurricane Sandy, however, were somewhat more focused than those conducted following Hurricane Katrina. We were particularly interested to learn about how faith-based groups and nonprofit organizations had been involved with response and recovery.

In addition, whereas the Hurricane Katrina interviews had been dispersed across many different communities affected by the disaster, the interviews after Hurricane Sandy were centered on the Rockaway Peninsula in New York (and specifically the communities of Bayswater and Far Rockaway). The Rockaway Peninsula and, specifically, Bayswater and Far Rockaway were selected for a few reasons. First, when considering areas that had been affected by Hurricane Sandy, we tried to avoid places that were predominantly second homes, or had a large seasonal population. This ruled out many areas in New Jersey as well as a few areas in New York. Second, we wanted to focus on a community that was not especially well-off in traditional terms (e.g., income level or educational level). According to the 2010 Census, Far Rockaway and Bayswater had a high school graduation rate of 73.8 percent, an average household income of $38,631, and a poverty rate of 27 percent.[30]

The research team conducted interviews in July and August of 2013 and conducted follow-up interviews in July 2014. In total, 16 interviews were completed. In addition to the interviews, we used secondary sources to find information and corroborate interview data, and also consulted written histories of the area as well as contemporary news stories.

In the Subsequent Chapters, We Describe What Various Entrepreneurs Did to Spur Recovery

The interviews that were conducted speak to the important roles entrepreneurs played after Hurricanes Katrina and Sandy. Interviews with entrepreneurs revealed the challenges they had to overcome as they attempted to start or reestablish their enterprises following the disasters. Interviews with disaster victims revealed the various ways in which they relied on these entrepreneurs during the rebuilding process. Drawing on qualitative data collected after the two disasters described above, we focus on entrepreneurs working after Hurricanes Katrina and Sandy to provide goods and services (chapter 5), restore and replace disrupted social networks (chapter 6), and signal that community rebound is likely and, in fact, under way (chapter 7). Each of the case studies presented focuses on a particular entrepreneur and highlights their background, their connection to the neighborhoods that they serve, and the specific function that they performed. The cases were selected so that we present a diversity of entrepreneurs performing a variety of tasks. Still, we do not pretend that, even taken together, they speak to all the things that entrepreneurs did to help spur community recovery. Furthermore, we will not distinguish between the different types of entrepreneurship in the cases that we examine in detail in the next few chapters. Recall, as we argued in chapter 2, that delineating between commercial, political, social, and ideological entrepreneurship overemphasizes the differences and underemphasizes the similarities of the different types of entrepreneurial action. We argue that any effort to describe these entrepreneurs as being of a certain type would be crude, at best, and would obscure more than it would illuminate.

Also, in the subsequent chapters we do not speak about entrepreneurial efforts that failed.[31] There are three reasons for this. First, the empirical record is silent on those entrepreneurial efforts that did not succeed (at least in some minimal way) in providing goods and services, restoring social networks, or signaling return (at least in the period that we were studying them). One of the reasons that we recommend ensuring there is a space for entrepreneurial experimentation after a disaster (see chapter 8) is because it is impossible to predict ex ante which entrepreneurial attempts are likely to succeed and which are likely to fail. Given that epistemic challenge, the correct posture is, arguably, to allow for entrepreneurial activity to occur, including efforts that may seem likely to fail. It is only through this process of discovery and experimentation, we argue, that entrepreneurs and other disaster victims will be able to overcome the collective action problem and go about recovery. This, of course, is true of entrepreneurial failure in all circumstances, including during mundane times as well as after disasters.

Second, we are silent about failures because we want to highlight the efforts of those entrepreneurs who succeeded in helping to bring about recovery and to stress the challenges that even these successful entrepreneurs had to overcome. There is a great deal of frustration and failure in the cases that we present, even if the entrepreneurs ultimately succeeded. For instance, many of the entrepreneurs we highlight ran into regulatory roadblocks, including Alice Craft-Kerney (discussed in chapter 5) who faced issues receiving the correct permits, which delayed the opening of her health clinic in the Lower Ninths Ward. Additionally, Doris Voitier (discussed in chapter 7) underestimated the demand for her services and had to make adjustments to satisfy demand.

Third, and perhaps most importantly, entrepreneurs who fail to perform one or more of the post-disaster entrepreneurial functions that we highlight might, nonetheless, succeed in performing other functions necessary for community rebound. The entrepreneur that tries and fails to start an enterprise, or organization, to deliver some good or service might still signal through his efforts that community recovery is under way. Similarly, despite her best intentions, an entrepreneur's enterprise (for example, her restaurant) might never develop into the social space that she wished it would become, but it may, nonetheless, provide necessary goods and services. The story of entrepreneurship after disasters is a story about what entrepreneurs can and did do, not a story about what they failed to do.

Chapter 5

Goods and Services Providers

As we have discussed in chapter 3, the post-disaster environment presents several unique challenges for disaster victims. Critically, as a result of the disaster, there is a sudden increase in the demand for certain goods and services, especially contracting services, building materials, and household items. As Chamlee-Wright (2010: 44) describes, "disaster sparks fierce demand for essential goods, services, and expertise the ordinary person does not possess." Moreover, there is a sudden disruption in the pre-disaster sources of these goods and services, since many of the enterprises that serviced the area prior to the disaster as well as many of the employees of those enterprises will also be affected by the disaster. Entrepreneurs recognize this increased demand and work to satisfy it by resuming pre-disaster operations, offering goods and services that they did not offer prior to the disaster, or reorienting pre-disaster enterprises and service offerings to meet the now heightened demand. In the language we introduced in chapter 3, entrepreneurs lower the cost of returning and rebuilding and/or increase the benefits associated with returning and rebuilding by restoring or providing new goods and services. For instance, contractors, carpenters, plumbers, and electricians enter the post-disaster market (often from nearby communities) or expand their activities and offerings, thus, making available the technical skills needed to repair and rebuild homes. Similarly, entrepreneurs create or reopen establishments that sell groceries, clothing, furniture, appliances, and other household goods, allowing victims to replace items that were lost during the disaster. Additionally, entrepreneurs create or reopen restaurants, day care centers, schools, gas stations, laundromats, pharmacies, and other enterprises that offer essential services.

Several scholars have drawn attention to the fact that certain goods have to be replaced as recovery gets underway, along with the continuing need to purchase everyday items and services after a disaster. A growing literature, for instance, argues that entrepreneurs continue to operate in hostile environments and may even be better able to cope with hostile environments than non-entrepreneurs (Bullough et al. 2013; Galbraith and Stiles 2006; Baron and Markman 2000; Branzei and Abdelnour 2010: 806; Joseph and Linley 2008; Markman et al. 2005; Tedeschi and Calhoun 2004). Indeed, perhaps the most obvious and least controversial role for entrepreneurs post-disaster is to provide goods and services. After all, entrepreneurs provide goods and services during mundane times and, thus, it is reasonable to believe that they will perform this function in other situations as well.

Entrepreneurs can work to provide needed goods and services through traditional commercial activities as well as by operating outside of commercial sectors.[1] As scholars have noted, nonprofit organizations do supply important goods and services to returning residents (Chamlee-Wright and Storr 2010a; Smith and Sutter 2013; Coyne and Lemke 2011). For instance, thousands of volunteers offered to help gut houses and clean debris following Hurricane Katrina, providing a critical service during the recovery process.[2] And, following the 2011 tornado in Joplin, Missouri, entrepreneurs helped to mobilize 92,000 registered volunteers and 749 church, charity, business, hospital, and school groups that participated in cleanup activities (Smith and Sutter 2013). These entrepreneurs accessed local knowledge to determine which goods and services were most needed and often relied on existing social networks to coordinate their efforts, in addition to sharing information related to the provision of goods and services with disaster victims (Chamlee-Wright and Storr 2009a, 2010a; Storr and Haeffele-Balch 2012).

Notably, the Hurricane Katrina victims we interviewed spoke of (1) the level of devastation following Hurricane Katrina and (2) the challenge presented by the increased need (at least in the short term) for certain goods and services as well as the certainly stretched and likely reduced number of suppliers of those goods and services.

Recall that Katrina caused over $100 billion in damage and over 100,000 homes in the Greater New Orleans Region were severely damaged or destroyed. In some communities, the damage was so severe that almost nothing could be salvaged. Parts of the Lower Ninth Ward, for instance, were completely destroyed. As one resident of the Lower Ninth Ward, Daisy Dubois†, described, "All of it's under water. I [saw] the [footage on] television, I cried for days. When your life savings is all watered down…people were devastated."[3] Other neighborhoods suffered several feet of flooding, as well as mud, mold, toxic material, and debris once the water receded. Matt Brown† of Broadmoor, for instance, described the mountains of garbage alongside the road on his street as neighbors gutted everything from their homes, which consisted of "sheetrock, roof material…and decorations…. Rugs, carpeting, ceiling fans, all kinds of damaged stuff."

In some cases, residents were not allowed back in their homes for several weeks. When people were finally allowed to return, they needed access to cleaning supplies, gasoline, and generators to start clearing out debris and begin the rebuilding process. Furthermore, grocery stores and gas stations were damaged and transportation was limited, resulting in food and gas shortages and the need for supplies to be brought in from outside the community. Entrepreneurs, however, provided goods and services that helped meet these needs.

Similarly, the Hurricane Sandy victims we interviewed discussed the damage (recall, over $60 billion in property damage) caused by the storm and the need to replace various household items. Again, some communities experienced more devastation than others. Several of the communities along the barrier islands in New Jersey, including Atlantic City, suffered severe damage. In Breezy Point on the Rockaway Peninsula in New York, the rising water touched an electrical line, setting fire to as many as 130 homes. Waves reached over 20 feet in some places, and many buildings in Bayswater and Far Rockaway had up to six feet of flooding. Residents lost furniture and appliances, clothing, and other personal items. As Rabbi Kruger of Bayswater describes, in the Jewish community in Bayswater and Far Rockaway, "We had 110–115 families that had significant damages." He further explained that the needs of disaster victims were varied, with some families needing heat and hot water, others needing to drain and repair their basements, and yet others needing more extensive repairs to fix significant structural damage. For instance, pumps were needed to remove floodwater, and generators were in high demand for the two weeks following the storm when the community was without electricity. One Bayswater resident, Tom Schmitz†, recalled that residents needed chain saws to cut through fallen trees and clear roadways. Luckily, he notes, there were many plumbers and electricians who lived in Bayswater that offered to help their neighbors. Without electricity, refrigerators could not preserve food, and access to grocery stores was a challenge as many cars were flooded and gasoline was in short supply. Entrepreneurs established food banks and organized donations and volunteers to ensure that affected residents had access to food during this time period.

The examples described here highlight how entrepreneurs provided needed goods and services after Hurricanes Katrina and Sandy.

Providing Furniture and Appliances in Chalmette

Mary Ann Patrick† is the owner of a furniture store in Chalmette in St. Bernard Parish. Patrick provided crucial resources during the rebuilding process after Katrina, including mobile homes, furniture, and appliances.

St. Bernard Parish is a predominantly white middle-income community east of Orleans Parish. Chamlee-Wright and Storr (2011b: 273) have described St. Bernard Parish as a "close-knit, family-oriented community comprised of hard workers." Prior to Katrina, the community was 84.3 percent white with a median household income of just under $36,000.[4] Compare this to Orleans Parish, which was only 26.6 percent white with a median household income of just over $26,000. Chalmette is the largest city in St. Bernard Parish, with a pre-Katrina population of 32,069.[5] Before the storm, the major employers included Exxon Mobil Chalmette Refinery, the Murphy Oil Refinery, Domino Sugar, Chalmette Medical Center, Boasso America Corporation, as well as several national retail outlets.

During Hurricane Katrina, parts of St. Bernard Parish experienced flooding as high as ten feet. Much of the community could only be accessed by boat or helicopter. And once the water receded, mud and other debris blocked roads, making it difficult for residents to return to their homes. As Doris Voitier, whose post-Katrina efforts are discussed in chapter 7, explained, people who had evacuated expected to be able to return to their homes in St. Bernard Parish in a few days. Instead, when they were finally able to return weeks and months later, "they had lost everything. I mean, everything was gone. They'd lost their homes. They'd lost every possession they had." Similarly, Nancy Carter† estimates that she had 18 inches of debris throughout the house, including mud and oil. Indeed, oil from Murphy Oil Refinery coated the neighborhood. Carter recalls seeing a ring of oil approximately one foot wide around her house. In addition to devastated homes, businesses, public buildings (such as schools), and other essential services (such as hospitals) were severely damaged.

As a consequence of the storm and its aftermath, 163 of St. Bernard Parish's residents died. Nearly the entire population of the parish was affected by flooding (i.e., 81 percent of owner- and renter-occupied housing units were damaged).[6] Businesses in St. Bernard Parish suffered over $500 million in losses, and there was over $3 billion in damage to property and personal belongings. Not surprisingly, the population in St. Bernard Parish three years after Katrina was only at 46 percent of pre-Katrina levels (see table 4.1). According to Chamlee-Wright and Storr (2011b: 274), however, "community members who returned within two years of Katrina tended to adopt a self-reliant strategy depending primarily on their own efforts and informal support from kin and other community members rather than formal support from external sources."

Patrick grew up in Chalmette and comes from a line of business owners. Her father started the business that she now runs and her grandfather owned a grocery store in New Orleans. She proudly stated that she has been a business owner for over 38 years at the time that we interviewed her in 2007. "I started on Broad Street in '69. So I've been in business, June 1st, for 38 years." Prior to Katrina, Patrick employed five or six people and sold mobile homes, furniture, and appliances.

Hurricane Katrina, and the flooding that followed, damaged and disrupted Patrick's business. The floodwaters were as high as eight feet in her large furniture showroom, and when she returned to view the damage after the storm, she recalled that "it was like walking on the moon, we had like... 12 inches of mud." She also lost three quarters of the concrete roof on one of her buildings as well as all of her inventory. Furthermore, her home was destroyed, forcing her to live in a trailer after the storm, and many of her family members were displaced. And, as she describes, "before the storm I had like 45 to 50 of my family members in this area. Now [two years after Katrina] I have like five... And after the storm... most of their homes had to be demolished."

Given the level of devastation in the community, Patrick recognized an opportunity (in the Kirznerian sense) to provide cleanup services as the recovery process got underway. Patrick recalled how she encouraged a family member who was struggling to find a way forward in the days following the storm. She recounts how she told him that "we're going to get up and go to work. We're going to . . . the Home Depot with my credit card and they got the shovels, and wheelbarrow, and [stuff] like that, and they started cleaning houses. And they were cleaning houses on a daily basis." As a result of their efforts, her family members were able to earn enough money to help them get back into their homes. To return to the language introduced in chapter 3, Patrick was able to alert them to an opportunity to help finance their recovery by providing a needed service to the community.

Approximately two months after Katrina, Patrick returned to the area. She needed a place to live, had to repair her home, and wanted to get to work on reopening her businesses. She had difficulty finding a place to temporarily relocate her business as she repaired her Chalmette location, but eventually she was able to lease a building in Slidell, a city in St. Tammany Parish, north of St. Bernard Parish. Thankfully, Patrick had building and business interruption insurance, which meant that she had enough money to lease the temporary location. However, she struggled to get the necessary permits and approvals. As she describes, "they wouldn't give me a permit to get my phones and the electricity and that on. They kept telling me the building had to have more things done to it and stuff like this. And I leased it, like, in October, and it was still November, and I still hadn't been able to get my business license and everything else." Finally, in December 2005 she opened the Slidell location. Patrick did not let the hurdles stop her, explaining that, "My dad taught me how to work."

In the meantime, Patrick was traveling back to Chalmette to clean and repair the buildings at her original location. She discovered that FEMA had determined that her buildings, as well as all the other buildings in the surrounding area, were condemned. Moreover, the Parish government had begun to plan their demolition. Her insurance company, however, disagreed with FEMA and the Parish's assessment. So did Patrick. She hired a structural engineer to assess the building. After several months of back and forth with the Parish, Patrick finally won—the Parish agreed that she could renovate the building, rather than having it demolished. As Patrick recalls, "It took me six months to convince the Parish that they weren't going to demolish my building."

Patrick recalls that the first time she was allowed to return to her damaged facility, one of her regular customers drove by and stopped to say hello. She told Patrick that "[N]ow that you're here I know that everything's going to be all right." As we will discuss further in chapter 7, by reopening her business, Patrick also served as a focal point, showing that she was committed to returning and signaling to others that the community was likely to rebound. Again, in the language we introduced earlier, Patrick not only lowered the cost that her fellow disaster victims would have to incur if they were to return and rebuild, by providing needed goods, but also increased the probability that her fellow disaster victims would, indeed, return.

Following the storm, Patrick also shifted the products she carried to accommodate what people needed most. A disaster victim herself, Patrick was alert to what people in her community and surrounding areas needed. For instance, she provided mattresses, washers, dryers, and other appliances that were in high demand during the recovery process. Indeed, having something to sleep on was often among residents' top priorities. Patrick also noted that customers were eager to purchase washers and dryers, which they hooked up outside their FEMA trailers. And her revenues from sales were positive feedback that she was providing the goods people desired.

Patrick reports that her business was successful in the post-Katrina environment, in part because she recognized the changing needs of her clientele and was able to adapt to the new circumstances. She was alert to, noticed, and, despite the challenges, pursued a profit opportunity by providing disaster victims the goods they needed to rebuild. "You just deal with it as it comes along," she explained, "You know, it's a situation that you just deal with it as it comes along." And, despite difficulties and challenges, Patrick also remained optimistic and confident her community would recover. She stated, "You know, I see a lot of people decided that, they didn't think they were coming back...have returned. [T]hey're seeing that they miss it...the first thing they say, this is my home and I want to come back." By providing needed goods and being committed to her community, Patrick helped drive recovery.[7]

By focusing on reopening their businesses and getting back to work, entrepreneurs can not only help themselves but also help their communities return to a sense of normalcy. The success of big-box stores, like Wal-Mart and Home Depot, in quickly providing needed supplies and encouraging residents and other business owners to return in the post-Katrina context is well documented (Horwitz 2009a; Rosegrant 2007a, 2007b). Furthermore, Chamlee-Wright and Storr (2008) emphasize the importance of local commercial entrepreneurs after Katrina, highlighting how they provided needed goods and services while also signaling a commitment to the community and its recovery. In the language introduced in chapter 3, entrepreneurs like Patrick lowered the costs of returning and signaled that recovery was under way, thereby aiding others in their calculus for deciding whether or not to return and rebuild their lives in Chalmette.

Repairing Damaged Roofs in and around New Orleans

Ed Williams† has lived in the Lower Ninth Ward all his life. Prior to Hurricane Katrina, Williams had his own roofing business, where he employed a few of his family members. After the storm, he helped rebuild his community.

Before Katrina, the Lower Ninth Ward had a population of approximately 14,000 people. The majority of the population (98 percent) was African American, the average household income was $37,894, and an estimated 36 percent of the population lived below the poverty line.[8] Growing up in the Lower Ninth Ward, Williams explained that it was a tight-knit community: "My mom and dad worked. I have a sister, it was fun. Everybody on my block was all, you know, all our parents worked. So we all had a, you know, we had a normal childhood. We wasn't rich but we wasn't poor. We was in the middle. If you wanted something you got it."

Williams recalled that his family had season tickets to the New Orleans Saints professional football team. He grew up with extended family nearby and also friends that were as close as family members. "The neighbors were like family... If your neighbors caught you doing something you would get your behind whipped. [If] your mama and daddy get home and find out, you get another one. That's the way it was when we was coming up." Speaking about his childhood best friend (who he is still very close with), Williams reminisced that "I call him a friend but he's more of a brother than a friend. Because I can go to his house and lay out with his mom and dad, and go in the room and we have, you know watch TV and do whatever I wanted to do. And the same way he could go to my mom and dad's house and do what he wanted to do."

Before Hurricane Katrina, over a third of individuals in the Lower Ninth Ward lived below the poverty line. And, while the City of New Orleans has one of the highest property and violent crime rates in the country, the Lower Ninth Ward was one of the worst neighborhoods in terms of crime and violence in the city (Landphair 2007). Criminologist Peter Scarf (ibid.) even referred to the Lower Ninth Ward as "the murder capital of the murder capital." Despite these stark statistics, interview subjects painted a very different picture of the community. Residents of the Lower Ninth Ward were proud to be from the neighborhood. In addition to descriptions of the social ills that plagued the community, interview subjects described the Lower Ninth Ward as having a rich history, tight-knit social networks, and high home ownership rates. Indeed, the Lower Ninth Ward had the highest home ownership rates in the city (approximately 60 percent) prior to the storm.

The Lower Ninth Ward suffered some of the worst damages following Hurricane Katrina. Levee breaches brought a 20 foot wall of water crashing into the community. The wave flattened houses, businesses, and anything else that stood in its way.[9] Much of the community experienced over ten feet of flooding. As one home owner, Raymond Thompson†, stated, his house had approximately six feet of water. He had to gut the walls and throw out everything in the house, including furniture, clothing, and personal items. Jim Evans†, who owned several homes in the Lower Ninth Ward, explained that one of his properties was reduced to "a pile of wood." Pointing to different places in the neighborhood, he said, "This house, that house was gone, the one next to it was gone and so we couldn't tell [which house] was what." And the emotional stress took a toll on many. Again, as Daisy Dubois described, "All of it's under water. When I seen the television, I cried for days. When your life savings is all watered down—and my mom and my dad—my dad's pressure was up. My daddy had a stroke." Williams was only able to keep the ceiling (which he disinfected two times) and salvaged two tubs and a toilet from his home.

Williams had not planned to evacuate. However, the day before Hurricane Katrina struck, he received a phone call from a cousin in Houston who urged him to gather up the family and leave before the storm hit. Williams and some 30 family members caravanned—six cars in total—to Houston. When they arrived, they all crammed into his cousin's townhouse. The next day, the cousin contacted her church to help find housing for the family.

A week after the Hurricane, Williams was able to sneak back into the neighborhood and examine his house. He was thrilled, the house had not been flooded. He planned to return to Houston, get a trailer, and drive back to the Lower Ninth Ward to retrieve furniture from his house. When he returned, however, the house was severely damaged. Rain had come through his wind-damaged roof, and when the city turned the water back on, the pipes in Williams' home had burst. Mold had then spread throughout the house. All of his furniture, clothing, and other belongings were destroyed.

Williams had experience with hurricanes prior to Katrina. For instance, when he was five years old, Hurricane Betsy hit New Orleans, resulting in major devastation across the Lower Ninth Ward, including flooding that reached the ceiling of many homes. The year before Hurricane Katrina hit, Williams and a few of his employees and family members traveled to Florida to repair roofs destroyed by a hurricane. Although he was no stranger to natural disasters, according to Williams, Hurricane Katrina brought more damage than he ever could have imagined. Williams returned to the Lower Ninth Ward approximately two months after the storm and began working around-the-clock repairing roofs in the area.

When he first returned, Williams went to St. Rose, a community west of New Orleans, where he had customers and friends that would help him get back on his feet. In fact, he had contacted one friend, Mark, when evacuating and asked for money. Mark had sent him several thousand dollars. Another resident, Mona, had offered to help Williams. He noted, "I said Mona, I'm coming back to go to work. About a day after that Mona gave me my first compressor that I had lost, all my tools I had lost, she gave me a generator, everything I needed to come to work with. Then when I got on my feet, when I came home, I started making a few dollars. I called her back and nobody wanted nothing back." In short order, Williams was working with eight family members repairing roofs. As more work came in, he hired a few extra hands.

Williams described his work as a contractor, saying that he offered customers a quality job, and that "I'm going to do what I can do to the best of my knowledge." Williams did not charge customers as much as other contractors did. He offered two explanations for his lower prices. First, he acknowledged that customers did not perhaps have the same level of confidence in Williams as they had in other contractors (because he had less experience than other contractors in the area) and, consequently, he had to offer bargain prices. Second, he stated that sometimes it took his crew a bit longer to finish a job. While a few potential customers sometimes opted for the higher priced contractor, others could not afford it, and Williams was able to fill that need.

Still, he reported that 95 percent of his customers were satisfied with his work. Williams saw himself as filling an important gap in the market, repairing roofs at discount prices.

Williams expressed several frustrations with federal disaster assistance programs. For instance, home owners would often hire Williams to repair their roof and then pay him with insurance money, savings, or, in some cases, Road Home Program money.[10] When there were delays in home owners receiving money, Williams would experience a delay in payment for his services. Typically, customers made several payments at different points in the process, meaning that any delays in payment often resulted in Williams having to operate with a negative cash flow for short periods of time.

It is clear that Williams has a strong connection to his family and the Lower Ninth Ward and was committed to return. He was also able to make a living and to employ family and friends by working as a contractor to repair roofs damaged by the storm. Furthermore, the experience of the storm solidified his commitment to family. On December 31, 2005, Williams and his long-term girlfriend married. He noted, "We was planning on getting married but we never done it. But when the hurricane came we said, 'what the hell. Give the family all something to do. We can party, get our mind off all our problems.' And, that's how we got married on the 31st of December."

As mentioned above, research has shown that entrepreneurs are important to post-disaster recovery (Chamlee-Wright and Storr 2008; Horwitz 2009a; Sobel and Leeson 2007). This is particularly true for entrepreneurs providing goods and services that aid in the rebuilding process, whether it is sheetrock, wood and nails from a hardware store, appliances from Patrick, or roof repair from Williams. Additionally, Chamlee-Wright and Storr (2014) argue that commercial activity influences and shapes social bonds that are necessary for recovery. Again, this is particularly true for Williams, who hired and trained family members and friends, and made valuable friendships with customers prior to the storm, who then gave him money and supplies to get back on his feet after the storm. In the language that we employed in chapter 3, entrepreneurs such as Williams lowered the costs of returning for disaster victims in the Lower Ninth Ward. Furthermore, his provision of contracting services, as well as his strong ties (bonding social capital) to the community and his efforts to encourage family and friends to return, signaled to others in the community that people were committed to recovery (increasing the probability that others would return).

Providing Health Care Services in the Lower Ninth Ward

Health care services are essential in both mundane times and following a disaster. As a registered nurse and health care provider in the city of New Orleans, Alice Craft-Kerney was aware of the health care needs of the New Orleans population. Following Hurricane Katrina, Craft-Kerney was inspired to help her community and opened a health clinic in the Lower Ninth Ward so that residents had better access to health care services.

Craft-Kerney had grown up in the Lower Ninth Ward and had strong ties to the community. Like Williams, she described it as a "family-oriented place." Craft-Kerney noted, "[W]hen we were kids growing up, I can remember sleeping by the front door with the fans and we didn't lock our doors. And, everybody knew each other. And, we had the type of community where we really couldn't get away with anything because that was Joe's mother and Joe's mother is going to tell my mother. And, there was true respect for the teachers. And, if the teachers... you did something wrong, you got it from the teachers, you got it at home, you got something from your parents because you shouldn't embarrass them." This strong sense of place was prevalent among the Lower Ninth Ward residents that we interviewed, with interview subjects reporting feeling tied to their community, its tradition, and its culture (Chamlee-Wright 2010; Chamlee-Wright and Storr 2009b).

Although she acknowledged that the community had its challenges, Craft-Kerney suggested that "the Lower Ninth Ward to me has gotten a bad name and the media has been spinning it the way they wanted." She also pointed out that community members were actively involved, explaining that, "We didn't have that problem like some people had in other areas of the city where PTA meetings, there was nobody there. We only had standing room always when we had PTA meetings. Because people were concerned about their children and there was always parental involvement."

Craft-Kerney did not evacuate in advance of Hurricane Katrina. She was working the evening before the storm and then sought shelter with a family member in another part of the city. When Craft-Kerney finally evacuated, she went to Albuquerque, New Mexico, and then later traveled to Houma, Louisiana, to stay with her sister. While in Houma, she heard news about the severe devastation in the Lower Ninth Ward. Recall that this neighborhood was one of the hardest hit.

In response, many community-based organizations stepped forward to help spur recovery and support residents who had lost everything in the storm. Craft-Kerney credited Common Ground, a relief organization that was formed after Katrina to spur recovery in the Lower Ninth Ward, as well as the efforts of churches and other volunteers.[11] In fact, Craft-Kerney took these acts of service and kindness as a sign from God that she needed to pursue her own act of service. Three separate events had a lasting effect on her. First, she had several huge fallen trees in her yard and needed someone to come with a chain saw and remove the debris. Several young men happened to come by and volunteered to help her with the trees. Not only did they help remove the trees, but they also refused any payment for their services. Second, a guest speaker at a local Baptist church had encouraged her to have trust in God and push away any doubts and concerns regarding the recovery process. And, third, Craft-Kerney had driven to a local distribution center to get drinking water. When she arrived, she saw that operations were temporarily suspended. A woman working at the center, however, saw Craft-Kerney and welcomed her in, and gave her cases of water and other donated items.

Craft-Kerney left with a car full of supplies and considered them as gifts from God. That day she resolved to open a health clinic in the Lower Ninth Ward to do her part to provide for the community.

She recalled that her mother was concerned about her decision and worried that she would have a hard time remaining financially secure. Craft-Kerney reminded her mother that the Lord says "you have to walk back by faith and not by sight... [and] be obedient." Craft-Kerney was determined to achieve her goal, stating that, "[I]n neighborhoods you need certain things in order to survive, for sustainability. You need schools. You need churches. You need medical care. You need places to shop."

She knew, however, that she would need help. She reached out to her friend Patricia Berryhill, who, in addition to over 30 years of health care experience, possessed what Craft-Kerney refers to as a "culture of excellence." Berryhill, who had recently remarried and moved to a new home, donated her old home in the Lower Ninth Ward to serve as the building for the clinic. Common Ground organized volunteers to gut the house, which was badly damaged from flooding. Local contractors and tradesmen took on the project, pro bono, installing new walls, electrical wires, and all the necessary components for the clinic. Home Depot donated $15,000, and organizations like Leaders Creating Change through Contribution raised $30,000 for the cause (Chamlee-Wright and Storr 2008). And, others donated medical supplies and resources for the clinic.

On the day that was supposed to be the grand opening of the clinic, however, the city shut down the effort because they did not have the proper permit. While they had a residential permit, they needed a commercial permit to provide health care services to the public. With the help of a local lawyer and an architect, Craft-Kerney was able to get all the inspections and paperwork completed and the clinic finally opened on February 27, 2007 (a year and a half after the storm).

Craft-Kerney instituted a policy that the clinic would not turn anyone away.[12] As a result, the clinic averaged approximately 80–85 patients each week. And she estimated that 90 percent of the clinic's patients did not have health insurance. Instead of needing to provide insurance, patients only had to pay $25 per visit, whether being treated for physical ailments or mental health issues.[13]

There are significant health disparities in New Orleans. For instance, the African American population (recall that the Lower Ninth Ward is 98 percent African American) in New Orleans is more likely to die of chronic illness compared to the white population.[14] Moreover, the city has some of the highest rates of infant mortality, chronic diseases (such as asthma and diabetes), and sexually transmitted diseases (such as HIV/AIDS). A Kaiser Family Foundation (2013) survey of New Orleans found that 41 percent of respondents reported they had hypertension, diabetes, asthma, or other breathing problems or another chronic health condition.[15] There are numerous factors that contribute to poor health outcomes, including inadequate access to health care services. The same Kaiser Family Foundation survey found that

27 percent of respondents did not have a regular source of health care aside from a hospital emergency room.

Craft-Kerney notes that Hurricane Katrina and her experiences following the storm changed her outlook about what is important and what sort of challenges a person is capable of overcoming. She concluded:

> You know at some point I guess when pre-Katrina, I thought I had it going and I had my house and I had my car and I had a little money in the bank and I was doing all right. I was cruising right along. And, after Katrina, I think what I realized is you know I don't need anything and I can still be happy. I can still have joy in my heart you know. I'm healthy and I have a sound mind. And, as long as I have that...I don't feel defeated. Because people will say, "this is a Katrina survivor." And I say, "no, I'm not a Katrina survivor. I'm a Katrina overcomer." I said because I know with the Lord I can overcome anything. It's not just about surviving. It's overcoming these obstacles and barriers and challenges that they're putting in your way.

Research has highlighted the importance of entrepreneurs in providing needed social services, such as health care (for instance, see Drucker 1985). Such entrepreneurs are also crucial in the post-disaster context. Chamlee-Wright and Storr (2008) examine how local entrepreneurs provided numerous goods and services after Hurricane Katrina, which encouraged and supported rebound. Craft-Kerney is a prime example. As they argue, "In the fall of 2006, she and a team of volunteers observed that, for the community to rebound, residents would require access to basic health-care services, and they determined to meet that need by establishing a health-care clinic" (ibid.: 7).

Craft-Kerney was inspired by the efforts of her neighbors in the Lower Ninth Ward after Hurricane Katrina and set about opening a health care clinic for residents. Her clinic provided affordable treatment for physical ailments as well as mental issues to residents whether or not they had insurance. In the language utilized in chapter 3, Craft-Kerney recognized an opportunity to provide a service in response to unmet needs in the Lower Ninth Ward, adding an additional benefit for displaced residents to consider as they decided whether or not to return and rebuild. Although the clinic closed in December 2010 because of cash flow problems and new Medicaid requirements regarding clinics to have chief financial officers, which Craft-Kerney could not afford (Barrow 2010), her presence in the community made a difference during the long recovery process.[16]

Providing Safety in Bayswater and Far Rockaway

In November 2011, Jason Shtundel organized the Rockaway Citizens Safety Patrol (RCSP), a volunteer group dedicated to ensuring the safety of community members in the Bayswater and Far Rockaway communities on the Rockaway Peninsula in Queens, New York. The RCSP was well regarded in the community before Hurricane Sandy; and following the storm, the group provided additional surveillance and assistance.

The Rockaway Peninsula in New York is extremely diverse on a range of measures—socioeconomic status, race and ethnicity, as well as educational attainment. Beginning in the mid-twentieth century, New York City officials sought to relocate poor minorities and selected the Rockaway Peninsula as a suitable location, because it was separated from New York City and officials did not anticipate the existing population on the peninsula to organize opposition (Kaplan and Kaplan 2003). Public housing developments were constructed in the Rockaways and within short order the peninsula contained over 50 percent of the housing projects in the Borough of Queens (while only containing 0.05 percent of the population) (ibid.: 3). While housing projects were in abundance, other public services, such as social work and mental health services, were in short supply and of poor quality (ibid.).

Before Hurricane Sandy, Bayswater and Far Rockaway had a population of around 60,000 and an average household income of just under $40,000. The community was racially diverse, with roughly 30 percent identifying as white, 50 percent as African American, and 25 percent as Hispanic or Latino.[17] There was also a large Orthodox Jewish population in the area. Indeed, as Kaplan and Kaplan (2003: 4) note, "Orthodox Jews, who set up separate schools and religious institutions, established a rapidly growing, tightly knit enclave, mainly in Far Rockaway and Bayswater." Indeed, today the Orthodox Jewish community on the Rockaway Peninsula has an array of club goods including schools, religious centers, social service programs, an ambulance service, and, as we discuss below, a civilian patrol.

Hurricane Sandy made landfall on October 29, 2012, bringing with it high winds and rain as well as a storm surge from Jamaica Bay that caused several feet of flooding in Far Rockaways and Bayswater. The community was without power for two weeks following the storm, during a time when temperatures began to dip below freezing. Many families lost furniture, appliances, and other personal items and sustained wind and water damage to their homes. Additionally, public transport was disrupted and gasoline shortages made entering and leaving the geographically isolated Rockaway Peninsula difficult in the weeks immediately following the storm.

While the water receded fairly quickly, the conditions and isolated nature of the community made mold an issue and increased the need for immediate assistance to pump out water and repair damaged homes and businesses. As Rabbi Kruger, a resident of Bayswater, noted, "Mold is insane. You can [have a] house that looks [like] it's in fine shape and tomorrow you come back and there's black all over the walls. It's amazing; and it races up, inside the walls. You have water this deep and then tomorrow you've got it in the ceiling of your second floor." And in response to these pressing needs, numerous organizations and community members stepped in to provide support and aid in the immediate days following the storm.

Prior to Sandy, civil society within the Orthodox Jewish community was thriving, and there were numerous providers of goods and services tailored to the specific needs of the community. For instance, the Hatzalah ambulance service is a private, volunteer-based ambulance service that works within the rules and practices of the religion, while also having faster arrival times than equivalent city services. Likewise, organizations like the RCSP emerged to deal with the needs and concerns of the community. The RCSP is one organization that stepped up their efforts following Hurricane Sandy.

Founder Jason Shtundel, who grew up in the area and has long been a part of the community, says that he started the group after he became frustrated with petty crime in the community. His car was broken into in the beginning of November 2011. After Shtundel repaired his car, it was broken into again a few weeks later. At that point Shtundel started to talk to friends and neighbors about starting a civilian patrol.

Although the 101st Precinct serves Bayswater and Far Rockaway, community members wanted additional protection against petty crime, including car break-ins and graffiti. In this regard, the RCSP is not unlike the other community volunteer patrols that exist throughout New York City. Within the extended Jewish community specifically, for instance, there are shomrim (Hebrew for "guard") groups, which are Jewish civilian patrols that have been organized to address burglary, vandalism, muggings, assault, and other crime in their local communities. The RCSP and shomrim groups keep their eyes and ears on the street. If a patrol member suspects foul play, they call the police. Patrol members can also monitor the changing location of a suspect and then communicate updated information to police. As Shtundel explains, "One of the benefits of having a citizen's patrol is that we know our neighbors and we know what is out of the ordinary. If I see a stranger in a car that I know belongs to my neighbor, I do not have to think twice about calling the police."[18]

The RCSP has two programs. The first is their night patrol, which is staffed by volunteers who commit to keeping watch in three-hour shifts. The volunteers receive a radio and flashlight and patrol either on foot or in a car. The second component is the 24-hour hotline, where residents can call to request assistance or report suspicious activity.

Unlike shomrim groups, the RCSP is a multiethnic organization, and has been so since its establishment in 2011. To get participation from the non-Jewish community, early RCSP volunteers attended a variety of civic associations and community council meetings. Shtundel estimates that 80 percent of volunteers are Jewish and 20 percent are non-Jewish.

The RCSP expanded their role in the community in the days before and after Sandy. Prior to Sandy, volunteers helped to evacuate sick and elderly individuals in the community. Following the storm, the RCSP set up a station at the Young Israel of Wavecrest and Bayswater synagogue, which had quickly become a relief and distribution center to assist with community needs. RCSP volunteers helped pump water out of flooded homes and clear fallen trees and debris, while also continuing to patrol the neighborhood and run the hotline.

For two weeks after the storm, the RCSP had volunteers out on the streets 24/7—extending patrol to daytime hours in addition to nighttime. While other areas on the peninsula had police and the National Guard on duty to help deter looting and limit property damage, Bayswater was patrolled primarily with RCSP volunteers and had no real issues with crime immediately following the storm.[19] And, in many of our interviews, residents mentioned the importance of the patrol following the storm. For instance, Rabbi Kruger recalled, "I don't think it was more than a couple of days before the local policemen station [sent] an officer [to] Young Israel to make sure those efforts were coordinated. They appreciated that we were covering an area that they could not have done as well as we did...And, that has made a huge difference in the general safety of the neighborhood."

Since then, the RCSP has expanded its patrol throughout much of the Rockaway Peninsula and has also talked to other groups, sharing best practices and experiences from the storm. Moving forward, the RCSP hopes to be approved as a Civilian Observation Patrol (COP), a program run by the New York Police Department.

Shtundel, who has a day job as a school principal and history teacher, recognized an opportunity to provide security services to his neighborhood and organized a volunteer effort to fulfill that need. After Hurricane Sandy, Shtundel and members of the RCSP noticed an opportunity to expand their efforts, increasing the number of patrols and helping with cleanup. Like Craft-Kerney, he provided a social service that was essential to relief and recovery efforts. In the language from chapter 3, Shtundel lowered the costs that disaster victims had to incur as they weathered the storm and went about recovery. His example further highlights the importance of entrepreneurs' ability to utilize their prior experiences and connections to identify areas for change and take actions that help their greater community (Chamlee-Wright and Storr 2008, 2010a, 2014).

Providing Direct Assistance to Residents in Far Rockaway and Lawrence

Rabbi Boruch B. Bender is the President and Founder of Achiezer Community Resource Center, a crisis center with offices in Far Rockaway and Lawrence, on the Rockaway Peninsula in New York. Achiezer was established in 2009 as a multifaceted support center for the Orthodox Jewish community, providing assistance with health, financial, mental, and other issues.[20] It is built around a network of local rabbis and synagogues, which works as both a referral and consultation network.

Rabbi Bender and his family are established members of the community. His father, Rabbi Yaakov Bender, runs the Yeshiva Darchei Torah, a prominent educational institute in the community, and instilled a sense of community service in all six of his children (Bensoussan 2012). When Rabbi Bender experienced a sudden illness that resulted in numerous consultations and eventually two surgeries, he realized that he could provide a service helping others to get through the grueling process of medical diagnosis, treatment, and recovery (Bensoussan 2012). Utilizing his experience and connections in the community, Rabbi Bender opened the crisis center. Within a few years' time, Achiezer grew into a one-stop shop to help community members navigate medical, financial, and legal processes.

Indeed, Achiezer became known as the place to turn to when one is faced with unexpected issues, whether medical problems or natural disasters. Their website is peppered with testimonials, such as "Thank you for providing a hospital bed, mattress, and [medical supplies] for my homebound father" and "There are no words to express our relief and our thanks for the unbelievable orchestration you arranged to move my grandmother to a hospital that could address her needs."[21] While conducting interviews, we noticed that when residents received phone calls about health issues or unexpected events, they would ask each other if they had talked to Rabbi Bender or asked him to help. Further, everyone we interviewed in the Jewish community either knew about, or was involved in, Achiezer's efforts after Sandy.

According to Rabbi Bender, the role for Achiezer to assist in disasters, and not just crises during mundane times, became evident during Hurricane Irene in 2011. As the storm was approaching, Achiezer received about 500 phone calls asking for help in preparing for the storm. He worked with Hatzalah, a local volunteer ambulance service, to transport 70 handicapped and elderly citizens to inland shelters (Bensoussan 2012). While Hurricane Irene did little damage to the area, it became clear that people would turn to Achiezer if a similar crisis arose. As Rabbi Bender remarked, "crisis infrastructure within our community... didn't exist," and so "people are going to call here, because there is nowhere else to call."

As Hurricane Sandy approached, many residents were skeptical about the seriousness of the storm, having experienced the inconvenience and minimal damage of Hurricane Irene the year before. However, to prepare as much as possible, Rabbi Bender held a meeting at Achiezer with community leaders, organization representatives, and local officials on Sunday afternoon. At the meeting, they discussed contingency plans in case the storm proved more powerful than expected.

Achiezer also issued email notifications with information about the storm, including road closures, evacuation procedures, reports on damage, and resources for response and recovery. They started out by utilizing their preexisting email database of roughly 9,000 contacts, and over 1,000 more people requested to be added to the list in the days following the storm. According to Rabbi Bender, the email notification system was "the most valuable service [Achiezer] provided... as the storm was hitting."

The next afternoon, as the storm came closer and as the weather got increasingly worse, the phone calls started coming in. That first night, Rabbi Bender estimates that they received 500 phone calls as residents dealt with flooding, power outages, and damage. The power and phone lines then went down at the Achiezer offices, so they moved the operation to Rabbi Bender's home, used a generator for electricity, and set up 16 phone lines in his dining room by forwarding Achiezer's hotline number to his home phone line (Bensoussan 2012).[22] Rabbi Bender recognized the importance of the preexisting hotline in ramping up efforts during Sandy. As he describes, "Now the crux of this unique infrastructure of having a one-stop shop, you know...one place to turn to, we were able to within hours have a hotline running, fielding thousands upon thousands of calls because of this different structure that was in place."

For the next week, Achiezer fielded approximately 1,500 phone calls a day and Rabbi Bender held meetings with community leaders and officials to begin planning for recovery. Within a few days of the storm, Achiezer helped transport over 300 families, whose houses had flooded, to temporarily relocate them to Brooklyn, Queens, and other locations (Bensoussan 2012). And once Rabbi Bender, his staff, and volunteers could move back into the office building, Achiezer also became a central location where residents could use their equipment and Internet connection to sign up for FEMA assistance. All of this happened before FEMA contacted Rabbi Bender and offered to help. As Rabbi Bender recalls, "We collected a bunch of laptops and iPads and set them up so that families with no computer access could come and sign up for FEMA help. About 130 families took advantage of the service" (Bensoussan 2012: 60).

The community's connections with other Orthodox Jewish communities across the country proved vital to getting resources. Three synagogues started getting donations and volunteers and became relief centers for the community, and Achiezer helped collect donations and secured generators and other supplies. Kosher food came in from Brooklyn, Queens, and upstate New York. Gas and generators were delivered from Baltimore. Trucks arrived to take damaged holy books to receive a proper burial, as required by religious doctrine and customs.

As donations started coming in and the focus shifted to recovery, Achiezer became the central location for receiving and distributing monetary donations to the community. To coordinate the influx of funds, Rabbi Bender, working with the Davis Memorial Fund, reinitiated the Community Assistance Fund (CAF) bank account, which had previously been used to help community members during the recession. They also developed a structure for assessing claims and distributing funds, which involved opening a separate bank account and recruiting local community members with needed expertise, including a board of trustees, attorneys, an accountant, and a professional fund-raiser.[23] The CAF team also enlisted the help of rabbis and other members of the community to work as representatives who would help community members apply for funding. The 48 representatives, spread throughout the Orthodox Jewish community on the Rockaway Peninsula, would work with people to fill out and submit applications, and then the board of trustees would review the applications and make final decisions on funding.[24]

Funding was broken down into three distinct phases. The first phase, called emergency cash assistance, was $2,000–3,000 per household for generators and emergency resources. Phase two, the coming-home project, averaged around $10,000 per household and went toward removing water and mold and other repairs, so that families could return to their homes as quickly as possible. And, finally, phase three provided major financial assistance for the rebuilding of homes damaged by the storm. Overall, over $11 million was raised, and CAF helped more than 1,000 families in the Orthodox Jewish community on the Rockaway Peninsula. Less than a year after the storm, Rabbi Bender expressed pride in his team's ability to raise and distribute the funds quickly and efficiently, "The staggering fact from this, which I am extremely proud of, and I want you to watch the media and the Attorney General speaking about the fact that a lot of places who raised money for Sandy, but it still didn't [get] out. We raised it, $11 million, and we gave out $11 million and there was no overhead costs."

The efforts of Rabbi Bender, Achiezer, and CAF were felt throughout the community. They were able to reach so much of the community through the deep connections of the fund's representatives. As Rabbi Kruger, the CAF representative for the Bayswater neighborhood, noted, trust was crucial to the quick and successful distribution of funds.[25] He explained, "Yes, there was also a level of trust; see, trust is very hard to get. People were very embarrassed that they were in need, very embarrassed, even though it was obviously not their fault. Sandy was very equal opportunity and nobody did anything wrong but people were very embarrassed. So trust and comfort and being willing to talk about your loses, it's not easy, so we became able to mobilize people who already had trust in the community, the rabbis, other community activists, whatever I had been doing, that definitely makes the whole thing easier."

Further, only a few people received FEMA assistance, but many received assistance from CAF. As one of the fund's representative noted, "I've been involved with a lot of families because I am the representative of the fund, the money that was raised on, I was one of the representatives to go into the houses and asked what they need and give it out. And, I did not speak to one person who got money from FEMA."[26]

By utilizing the preexisting organization and networks of the community, Rabbi Bender turned Achiezer into a disaster crisis center that funneled and distributed needed information and resources. Under Rabbi Bender's leadership, Achiezer shared information before, during, and after the storm through mass email notifications and the telephone hotline.[27] And Achiezer raised millions of dollars and distributed it quickly to members of the community. As Rabbi Kruger recalled, "Well, Achiezer was the nexus of everything and there were regular meetings there to get together and talk about what has happened, what can we do next, what are going to be the guidelines for the funds that are available." Rabbi Bender and Achiezer utilized both the strong ties (bonding social capital) of the Orthodox Jewish community in Rockaway and the weak ties (bridging and linking social capital) of the broader community to serve its members and drive recovery after Hurricane Sandy.

Research has examined both the importance and challenges of mutual assistance and charitable donations for post-disaster recovery. While the federal government is often looked to as the source of funding after a disaster, it is often local, community-based organizations that are effective in assessing the needs of the community and allocating the appropriate resources for relief and recovery (Sobel and Leeson 2007). For instance, Bolin and Stanford (1998) and Shaw and Goda (2004) found that community-based organizations provided direct assistance to victims of the 1994 Northridge, California, earthquake and the 1995 Kobe, Japan, earthquake, respectively. Skarbek (2014) has also described how the Chicago Relief and Aid Society, a voluntary association comprised of community stakeholders, managed contributions from individuals, businesses, and municipal and foreign governments and effectively delivered bottom-up, decentralized disaster relief to victims of the Chicago Fire of 1871 (see also Schaeffer and Kashdan 2010). Similarly, Storr and Haeffele-Balch (2012) argued that the Broadmoor Improvement Association in New Orleans was able to use its network and infrastructure to prove the vitality of their community, raise funds, and promote recovery after Hurricane Katrina. Likewise, Achiezer was able to raise and allocate millions of dollars in a quick and effective manner, ensuring that the community had the resources it needed to rebound. In the language from chapter 3, Rabbi Bender's efforts significantly lowered the costs of rebuilding that disaster victims had to incur, by gathering and distributing recovery aid through CAF.

In Summary, Entrepreneurs Provide Needed Goods and Services after Disasters

In the aftermath of Hurricanes Katrina and Sandy, entrepreneurs were able to provide goods and services that made it possible for residents to return and begin the cleanup, rebuilding, and recovery process. Even in the communities that were most severely impacted, entrepreneurs provided needed food, health care, security, and construction services.

Mary Ann Patrick, for instance, knew that reopening her furniture store would not only help her get back on her feet but also help her community in Chalmette rebound. She not only had the experience and connections to acquire and provide appliances and furniture but also had the ability to shift her product line to better suit her customers' needs following the storm. Similarly, as a contractor, Ed Williams was well positioned to provide assistance following Hurricane Katrina. Although his community in the Lower Ninth Ward was devastated, he was able to rely on the emotional and financial support of friends and family to return to New Orleans and begin work repairing roofs. Furthermore, Alice Craft-Kerney was inspired by the work of others in the Lower Ninth Ward and went about opening an accessible and affordable health clinic to better serve her neighbors. And it was her background as a registered nurse and her connections in her community that enabled her to provide a needed service.

Entrepreneurs engaged in similar efforts after Hurricane Sandy. Prior to the storm, Jason Shtundel perceived a problem in his community—the persistence of petty crime—and founded the Rockaway Citizens Safety Patrol in an effort to reduce petty crime and help his neighbors feel more secure in their community. It was a natural progression, then, that after Hurricane Sandy, the organization would expand its efforts to ensure the safety of community members so that other residents could focus on recovery. Similarly, Rabbi Bender's crisis center, Achiezer, was developed to address health and mental crises in mundane times, and evolved to handle crises during Hurricanes Irene and Sandy. Achiezer not only handled phone calls and transmitted informational emails after the storm but also became the hub for donations that were distributed swiftly throughout the community.

Chapter 6

Regrowing Uprooted Social Networks

In addition to providing needed goods and services after a disaster, entrepreneurs repurpose existing social networks and help community members restore and replace social networks that were disrupted by the disaster. Recall that social networks play a key role in helping individuals recover from disasters but that disasters can disrupt and destroy those networks. For disaster victims, social networks can be a source of financial resources, emotional support, mutual assistance, and information about how to navigate the challenges of the post-disaster environment. In the language that we introduced in chapter 3, entrepreneurs that restore disrupted social networks and facilitate the creation of new social networks increase the probability that others will return and lower the cost of returning by giving the victims access to mutual assistance.

People who are part of a disaster victim's social networks, especially her strong ties, are often disaster victims themselves and are often displaced by the disaster as well; their capacity to render assistance is, therefore, reduced. In such scenarios, entrepreneurs can discover and create opportunities for displaced residents to reconnect with one another, share information, and work together to bring about community rebound. As Chamlee-Wright (2010: 43) writes, "as difficult as the physical tasks of recovery are, these challenges seem almost easy compared to rebuilding the fabric of human relationships that make a collection of residential and commercial buildings a functioning community." Additionally, "the heterogeneous nature of social capital suggests that the reconfiguration and redirection of social capital resources is not a passive or automatic process but instead requires active and creative minds to tease out solutions" (ibid.: 43). Commercial and civic leaders are especially well suited to help community members restore and recombine social capital. As Chamlee-Wright (ibid.: 49) writes, "because of their particular tie to a neighborhood, businesses can play a direct role in reweaving place-based social capital, giving people greater reason to endure the hardships of the rebuilding process." This role is similar to, but distinct from, their acting as a focal point (discussed in chapter 7). While acting as a focal point, entrepreneurs coordinate recovery efforts and rally others to return or rebuild. An entrepreneur qua restorer of social networks primarily facilitates communication to connect dispersed residents with one another and with those who might be able to provide assistance. It is important to note that these roles are not mutually exclusive, and often one entrepreneur or a group of residents will act as a focal point for recovery by restoring a social network to connect dispersed residents or to spread information. Stated another way, social capital regrouping does signal community rebound that inspires others to engage in recovery, but it is a driver of community rebound in its own right.

Additionally, as individuals who frequently have weak connections to many different groups, entrepreneurs are uniquely situated to fill in structural holes or utilize linking social capital (Burt 1992, 1998). Similarly, entrepreneurs can connect members of different subgroups of a social network that were separated by the disaster. Entrepreneurs also work to connect aid suppliers (i.e., donors and volunteers) with aid demanders (i.e., disaster victims). Moreover, entrepreneurs can create social spaces where disaster victims can form and reform their social bonds. Stated another way, entrepreneurs can create opportunities for community members to spend time together "in ways their routine lives didn't afford and [help] to repair the social fabric of community" (Chamlee-Wright 2010: 48).

Other scholars have drawn attention to the disruption of social networks that can occur after disasters and the importance of disaster victims reconnecting their disrupted social networks to spur community rebound. For instance, as Holcombe (2007: 108) describes, "a disaster can have a devastating impact on a community by disrupting normal social functions and altering social relationships." See also Erikson (1976) and Quarantelli (1978). Similarly, Dynes (2006) and Adler (2010) have found that disasters change the pattern of, and obligations within, social networks. After a disaster, networks reorient around recovery. Also, pre-disaster obligations become less important in comparison to the needs and circumstances after the disaster. Communities with strong ties, such as the Vietnamese community in New Orleans East, have been found to recover quickly, relying on their close connections (Chamlee-Wright 2010; Chamlee-Wright and Storr 2009c, 2010c, 2011b; Aldrich 2011a, 2011b, 2012; Hurlbert et al. 2000, 2001). Additionally, communities that utilize their preexisting reservoir of weak ties and community organizations to reconnect and work toward shared recovery goals can spur recovery (Bolin and Stanford 1998; Chamlee-Wright and Storr 2010a; Murphy 2007; Shaw and Goda 2004; Storr and Haeffele-Balch 2012).

Notably, the Hurricane Katrina victims that we interviewed spoke about (1) the challenge of locating and communicating with similarly displaced disaster victims in their social circle (even close friends and family members) after they evacuated the area and were in exile following the hurricane and (2) the critical role commercial and civic leaders played in facilitating their reconnecting with old ties and forming new bonds.[1]

Recall that Katrina displaced over 400,000 residents from the region (Geaghan 2011). Moreover, the evacuation process that was adopted before and after Katrina exacerbated the disruption of social capital that can occur after disasters. As Weber and Peek (2012: 11) explain, "Tens of thousands of...evacuees had little or no choice in where they ended up after the storm...The ad hoc evacuation, combined with the lack of a central database for identifying and tracking evacuees' whereabouts, resulted in government-induced separation of family members."[2] Additionally, some victims were not even given an opportunity to visit their homes and assess the damage for several months following Katrina. For instance, residents from the New Orleans East community of the Mary Queen of Vietnam Catholic Church were scattered after Katrina, with some in Louisiana, Texas, Georgia, and Arkansas. To check in on everyone, their pastor, Father Vien (more details discussed in chapter 7), traveled back and forth between locations, taking photos, keeping everyone informed, and encouraging them to return. As he noted, "I was traveling in between...all of them in those locations and tried to bring them back into one area...[and]...I kept each community appraised." Likewise, displaced residents of the Broadmoor neighborhood were kept informed through text messages and chains of phone calls. One resident, Matt Brown, recalled that he got in touch with people via phone, albeit slowly. He said, "Well, with the cell phone, I think like about two in the morning, we might be able to get through to one or two people, and that's all that took, with just one, and once we made one connection, we just started a list. Who's number do you have? I got so and so. Like that, we just kind of started and then, as the crisis began to subside, we were able to make phone calls, but the text messages really helped tremendously."

Although Hurricane Sandy did not displace residents in similar proportions, the Sandy victims we interviewed did emphasize the importance of being connected to aid suppliers, and the important role commercial and civic leaders played in supporting the creation and maintenance of social relationships. For instance, Rabbi Bender and Achiezer (discussed in chapter 5) fielded over a thousand phone calls a day and emailed thousands of people information, both leading up to and following the storm, as well as served as a hub for collecting and distributing donations for the community. As a rabbi in Lawrence (a neighborhood near Far Rockaway on the Rockaway Peninsula) recalled, "Rabbi Bender's organization was coordinating, you know that whole effort...I would say within 24 hours of the hurricane, I got a call from Rabbi Bender asking if they can use this place, I would say 24 to 48 hours, if they can use this place for a distribution center for clothing."[3] In all, three synagogues became relief and distribution centers—one in Bayswater, one in Far Rockaway, and one in Lawrence—where community leaders set up patrol stations, food and supply pantries, and doctor's offices, thus facilitating rebuilding and recovery.

The following examples highlight how entrepreneurs restored social networks to spur recovery after Hurricanes Katrina and Sandy.

Proving Viability and Spurring Neighborhood Return in Broadmoor

LaToya Cantrell was the president of the Broadmoor Improvement Association (BIA) in the Broadmoor neighborhood in central New Orleans prior to Katrina and through her community's recovery from Hurricane Katrina. The BIA aims to enhance the connection and cohesion between racially and economically diverse residents and generally improve the quality of life in Broadmoor.[4]

Broadmoor is a historically diverse neighborhood, in terms of wealth, race, and ethnicity, that has been described as a "microcosm" of the larger city of New Orleans (Chamlee-Wright and Storr 2009c). Before Katrina, households in Broadmoor earned incomes ranging from $10,000 to $200,000 a year; very wealthy families lived just a few streets away from struggling single-parent families.[5] Median income for the neighborhood was just above $27,000, and roughly 32 percent of Broadmoor's residents lived in poverty before Katrina. Prior to the storm, the racial and ethnic makeup of the neighborhood was 68 percent African American, 26 percent white, and just below 4 percent Hispanic or Latino. In the 1950s, the neighborhood also saw an influx of Jewish residents who moved into the area in congruence with construction of the Chevra Thilim Synagogue.

The diversity within Broadmoor often resulted in an atmosphere similar to a small city rather than a typical New Orleans neighborhood. Still, a neighborhood organization was established in the 1930s (and was incorporated in the 1970s) providing a forum to discuss and lobby for public services and infrastructure as well as implementing a neighborhood watch program (BIA 2008).[6] The organization also spearheaded a fight in the 1970s against "blockbusting," a maneuver by the real estate industry to encourage the racial separation of neighborhoods, in order to preserve their "well-established, multi-racial/multi-ethnic community [that was] already living in harmony" (ibid.).

Cantrell was elected President of the BIA at the end of 2004, after joining the board in 2003. She first moved to New Orleans from Los Angeles, California, in the early 1990s to attend Xavier University, where she earned a bachelor's degree in sociology and minored in political science. She and her husband moved to Broadmoor in 2000, where they still live, along with their daughter. While president of the BIA, Cantrell was also the manager at the Greater New Orleans Education Foundation, where she helped public schools get access to private sector resources (Etheridge et al. 2006).[7] In 2012, and again in 2014, she was elected as the councilmember for District B of the New Orleans City Council, which covers the Broadmoor neighborhood.[8]

The Broadmoor neighborhood, centrally located in New Orleans, suffered major damage during the storm and flooding, with an average of eight feet of floodwater throughout the neighborhood.

Maggie Carroll, a resident of Broadmoor and the BIA secretary during the recovery, explained how her house was destroyed: "We had about three feet of water that sat for maybe ten days. Pretty much everything was destroyed, except we were able to save some art on the walls, which was pretty meaningful for us. We evacuated with our files and stuff, so we were okay, like family photographs and things like that. We took our pets, but the house was pretty much destroyed." Another resident recalled, "A lot of people had water pouring down their walls. If you didn't get water upstairs, you got water pouring down your walls. Your walls melted. Your floors rotted... By the time I added up what I got from the wind damage and what I added up from the total flood damage was more than I paid for the house."[9] And, Cantrell summarized the widespread effects by noting, "[E]verybody in this neighborhood was affected. Every property was damaged, all 2,400. The neighborhood was not glistening in any way, in no area."

The residents of Broadmoor had to wait weeks before they were allowed to return and assess the damage to their homes. As with other areas in New Orleans that received significant flooding, by the end of 2005, few residents had returned to Broadmoor (Chamlee-Wright and Storr 2009c). In early 2006, the Bring New Orleans Back (BNOB) Commission released its plan for the reconstruction of New Orleans. The plan included a proposal to turn the neighborhood of Broadmoor into a green space.[10] BNOB provided maps of the proposed plan, released in the *Times-Picayune* on January 1, 2006, which portrayed the area of Broadmoor as a "green dot." The commission stated that it would follow through with its plans unless communities could prove viability—that at least 50 percent of residents were committed to return—within four months. The plan shocked residents, who rallied together to save their neighborhood. They accomplished this through the leadership of Cantrell and the preexisting structure of the BIA.

Under Cantrell's leadership, the BIA utilized its existing neighborhood records and member lists to contact displaced citizens via text messages, emails, phone calls, and flyers and encouraged them to return to the neighborhood to discuss how to prove viability and begin actions toward recovery. These efforts resulted in a meeting held one week after the BNOB plan was released, which drew a large number of residents. At the end of the meeting, numerous plans had been put into action, including a petition against the "green dot," the creation of a BIA website to assist in outreach and provide a forum for residents to exchange information, and the creation of their own Revitalization Committee to bring planning and rebuilding under the community's control (BIA 2006).

At the meeting, Cantrell emphasized that residents needed to contact their neighbors and report to the BIA whether or not they planned to return. The BIA would then record the status of each property to document the number of residents who planned to return and, thus, prove viability of the neighborhood. Additionally, they used property tax assessor information and the National Change of Address Registry to locate displaced residents through a direct mail campaign.

Maggie Carroll attributed the large-scale mobilization effort to the already organized structure of the BIA that aided communication and outreach efforts. Additionally, the BIA's ability to form partnerships with the community's religious organizations, proactive residents, business owners, local and national nonprofits, and universities helped raise resources and to signal that Broadmoor was focused on recovery. The BIA used members' homes and local churches as impromptu headquarters for operations. Specifically, the Episcopal Church of the Annunciation, the Broadmoor Presbyterian Church, and Saint Matthias Catholic Church provided spaces for meetings and administrative support. These partnerships showed a united front for neighborhood recovery.

Further, the BIA used the skills and tools available in the diverse community to prove viability, highlighting the ability of private citizens to effectively tap into dispersed knowledge and use it to their advantage during recovery. Maggie Carroll articulated this nicely, "We've realized that we have such capacity here, and it already exists. People have so much expertise, and we're just able to really hone in and use those skills for the betterment of the entire neighborhood." For example, resident and BIA board member Virginia Saussy, who had advertising experience from working in the jewelry industry, developed a plan for a marketing campaign and worked with a Boston-based agency, Digitas Media, Inc., to take on the effort pro bono. Signs reading "Broadmoor Lives—In the Heart of New Orleans" were posted all over the neighborhood, and BIA members and residents spread the word of recovery and policy resistance through the local and national media (Warner 2006; Warner and Darcé 2006; Winkler-Schmit 2006). For instance, Broadmoor's story was told in a diverse variety of media outlets, including the New Orleans *Times-Picayune*, *National Public Radio* programs, and the Delta Airlines in-flight magazine *Delta Sky*.[11] By reconnecting with displaced neighbors, tapping the skills of members of the community, and rebranding the community as one of activity and improvement, Cantrell and the BIA were able to signal a commitment to community repopulation and revitalization.

Cantrell and the residents of Broadmoor were able to funnel resources toward reviving the neighborhood, including donations from the Clinton Global Initiative, the Carnegie Corporation of New York, Mercy Corps, and the Surdna Foundation, which went toward a charter school, a library, and a community center.[12] They also received assistance from the Belfer Center for Science and International Affairs at Harvard University's Kennedy School of Government.[13] Together, the BIA and the Belfer Center initiated a community-based development plan and set out mapping the recovery of Broadmoor. Their efforts were essential in proving the neighborhood's viability; as of January 2009, the Belfer Center had documented that 82.2 percent of the properties in Broadmoor had been rebuilt or were under construction (BIA 2009). The name recognition of Harvard alongside clear evidence of repopulation convinced donors to fund recovery and convinced politicians that Broadmoor was a viable community (and not destined to become a "green dot"). As a reward for its efforts, Broadmoor's community development plan was fully incorporated into all planning efforts after the initial BNOB plan.

The diversity in the neighborhood was utilized to bring together different skills, such as marketing and fund-raising, that helped ensure the viability and rebound of Broadmoor. In Cantrell's words, "I mean, our diversity is our strength." This highlights the importance of bridging and linking social capital in connecting diverse communities after a disaster. While the success of Broadmoor resulted from cooperation of the entire community, Cantrell provided the leadership needed to reorient the BIA's focus toward recovery and to keep the focus on community rebound. Not surprisingly, she has been recognized for her efforts, having received awards from the National Trust of Historic Preservation, Young Leadership Council, and others.[14]

Through the preexisting structure and leadership of the BIA, residents and business owners were able to communicate with one another and form a unified voice for proving viability, accessing government and nonprofit services, and working toward recovery. Returning to the language we introduced in chapter 3, Cantrell, through her efforts with the BIA, lowered the cost of returning for community members and increased the probability that others would also return by contacting and encouraging displaced residents to commit to returning. By acting quickly, Cantrell and the BIA were able to restore connections between community members and utilize the unique skills of residents of the neighborhood. They used emails, phone calls, direct-mailings, a website, and a door-to-door campaign to find dispersed residents, survey their desires to return, and exchange information about recovery. These efforts resulted in a strong rebound that ensured Broadmoor would be included in official recovery plans, receive funding, and become an active voice in the planning and redevelopment of New Orleans.

Several scholars have examined the Broadmoor recovery effort. For instance, Chamlee-Wright and Storr (2009c) and Chamlee-Wright (2010) argue that in the face of the "dreaded green dot," Broadmoor was able to overcome the post-disaster collective action problem and spur recovery. When faced with the potential elimination of the neighborhood, residents were able to refill the civil society vacuum that was created after the storm; they acted to return and rebuild in spite of the many burdens they faced. Chamlee-Wright and Storr (2010a, 2011b) highlight how Cantrell and the BIA played a crucial role in advocating for the community and spurring recovery. Similarly, Storr and Haeffele-Balch (2012) examines the heterogeneous and local nature of successful recovery in Broadmoor, noting how community-based organizations (like the BIA) can promote return and rebuilding in diverse, loosely connected neighborhoods. Similarly, Irazábal and Neville (2007) and Weil (2010) utilize Broadmoor as an example of how community and grassroots organizations can influence the post-disaster planning and recovery process.

Creating a Needed Social Space in Chalmette

When Ben Cicek, a resident of St. Bernard Parish since 2000, returned after being displaced for several months after Katrina, he noticed that there were few places open where people could get food, coffee, and Internet services. To fill this need, Cicek opened Da Parish Coffee House and, in so doing, created a needed social space in Chalmette, the parish seat of St. Bernard Parish.[15]

Recall that Hurricane Katrina and the flooding that followed devastated St. Bernard Parish. As much as ten feet of floodwater surged into the parish. Residents were prohibited from returning for several weeks after the storm, which only exacerbated the collective action problem associated with return. Additionally, Murphy Oil had a tank rupture during the storm, which, combined with damaged vehicles, meant that there was also oil residue throughout the parish. Despite the damage, many parish residents adopted a self-reliant recovery strategy (Chamlee-Wright 2010). As one resident, who had weathered out the storm and later left via boat, declared, "when you ride in a boat in a place, you grew up in all your life and you see it just devastated, the only thing you think of is there's nothing I can do but go on with it."[16]

Cicek had moved to St. Bernard Parish in 2000 from Turkey. Having decided to immigrate to America, he contacted a friend of a friend who lived in the French Quarter and made plans to move to the New Orleans area. When Hurricane Katrina hit, his rental house in Chalmette was destroyed. He then spent four months in Jackson, Mississippi, and one month visiting family in Europe before returning home. Since he lost his house in Chalmette, he found residence in the Uptown neighborhood in New Orleans.

While visiting a friend in Chalmette, he noticed an opportunity to open a coffee shop and fulfill a need in the community. His friend owned a restaurant and would frequently be asked if he also served coffee. Cicek had managed a coffee shop in Uptown for two years before Katrina, so they joined forces to start a new venture. They did not receive any loans from the Small Business Administration; instead they relied on their own savings, totaling $25,000. They rented space across the street from the restaurant, located at a major intersection in town, and did much of the renovations themselves, keeping it simple at first and improving as they went along. The shop served coffee and pastries as well as breakfast and lunch items, and offered free wireless Internet access to customers. And, while his partner later left the business, Cicek continued to invest and grow the business.

Cicek started the coffee shop because he realized that returning residents, contractors, and others engaged in the rebuilding effort could benefit from another place to eat and socialize. When asked what motivated him to take on the risk of starting a new business venture in the midst of the recovery effort, his response referenced both himself and the community: "I think it's just my personality. I took a risk and came to the United States seven years ago. I had only a few words in English... I would say that it's just personality. I like to take a risk and help the community a little bit and make money."

Cicek's customers varied in age, income level, and occupation, representing the heterogeneous population of St. Bernard Parish. While in the coffee shop, customers often talked to Cicek and others about what was going on in their lives, and the conversation was often about home owner's insurance, FEMA applications, and other challenges. People would share stories and experiences, exchanging information. "Everybody has different problems," he remarked, "That's what, they talk, we talk, I listen."

Further, Cicek explained how customers have become his friends and how he got to witness the community rebounding, "My first year I had a lot of workers from out of state or construction people. We had very few local customers. Now they're gone, so mostly, I have local customers. They have a lot of problems, you know, but they're coming back. They're fighting, or so I see it. We just talk about it everyday, so it's good to be there [for them]. It's an experience for me also to watch the development of the new city. It's sometimes exciting."

Cicek noticed a profit opportunity in opening a coffee shop, and not only provided a needed good, but also provided a needed social space for the community.[17] As Chamlee-Wright and Storr (2008) conclude, "since [Cicek] launched his business, a steady flow of customers enjoy both a fine cup of coffee and the restoration of market [and social] space." In the language from chapter 3, Cicek's coffee shop was a new benefit that disaster victims could introduce into their calculus to return. As residents returned and rebuilt their homes, they could stop by the Da Parish Coffee House and grab a bite to eat and a cup of coffee, as well as talk to one another about their problems, share information, and rebuild their social ties to the community. In other words, Cicek provided a social space in St. Bernard Parish where returning residents could restore disrupted social networks and facilitated their accessing the resources embedded in their social networks.

Entrepreneurs can play an important role in opening businesses that also facilitate the creation and reconnection of social networks after disasters. Chamlee-Wright and Storr (2014), for instance, examine how commercial activity fosters the creation and maintenance of social bonds and creates social spaces that encourage post-disaster recovery. Specifically, they argue that commercial relationships can enable recovery by allowing community members to "(1) provide and receive direct assistance to one another, (2) reconnect to pre-existing social networks, (3) establish new social and economic ties, and (4) overcome pessimism regarding the recovery effort by directly witnessing and gauging the pace of community rebound" (ibid.: 657). Similarly, Chamlee-Wright (2010) argues that commercial activity plays an important signaling role in the post-disaster contexts, not only by allocating resources to their highest valued use through the pursuit of profit opportunities but also by showing a commitment to return and facilitating social interaction. As she (ibid.: 51) explains, "the market is not merely a cheerleader for social capital: it is a principal provider of social capital. Cafes, bars, and restaurants, for example, provide an opportunity to reconnect old ties and create new ones." Cicek and Da Parish Coffee House created such a social space after Hurricane Katrina. By pursuing a commercial opportunity, Cicek enabled residents of Chalmette to reconnect and build new social networks that are necessary for recovery.

Reconnecting with Customers in Gentilly

Across town, Sandra Green† similarly filled a need for a social space after Hurricane Katrina by reopening a women's clothing store in Gentilly, Orleans Parish. Once it reopened, the store provided women in the surrounding neighborhoods with a space where they could experience a sense of normalcy and reconnect with friends.

Gentilly is a diverse, middle-class area in the northern section of New Orleans, next to Lake Pontchartrain.[18] Much of the area had average household incomes above those of the greater Orleans Parish. However, the clothing store is located in Gentilly Terrace, which has demographic and income statistics similar to the parish, with average household incomes of $42,000 (compared to $58,000 in the Fillmore neighborhood of Gentilly and $43,000 in Orleans Parish generally).[19] Hurricane Katrina brought roughly eight feet of floodwater and strong wind damage to Gentilly.

Green is the general manager of the clothing store, which she helped open after working as an assistant manager at another New Orleans location. Green's job as a general manager, as fulfilling the business activities of the retail chain, is not what makes her an entrepreneur (in the sense that we mean it here). Instead, it is the way she utilized her position to foster a social space for her customers, to reestablish social connections, that makes her an entrepreneur. In all, Green has worked for the retail chain since 1990 and has run the Gentilly store since 1993. Her store caters to African American women, ranging from teenagers to the elderly in the Gentilly, Desire, and New Orleans East neighborhoods. Much of the clientele are regular shoppers who visit the store not only to purchase clothing but also to talk with Green and each other.

Green grew up in Gentilly and both her grandparents were from the area. Once her father retired from the military in 1975, he moved the family to New Orleans East. And, when she got married, she and her husband found a place near them, where they now raise their two sons and one daughter. Green spoke of the importance of familial ties in the greater New Orleans area, noting that "it's really a family based area, it's generational. If one family buys in, more than likely the family stays."[20]

Green and her family, who had evacuated the Sunday before Hurricane Katrina hit, were displaced for two months following the storm. She and her immediate family stayed with her brother-in-law just 45 minutes away from New Orleans, in Napoleonville, while her other family members went to Texas. She returned in December and began the process of repairing her home, which had received severe wind damage but little flooding. Throughout this period, the retail chain continued to pay Green and the other employees. Because she continued to receive her salary, Green was able to focus on recovery without worrying about employment.

Her clothing store took over a year to reopen, which involved filing for, and waiting to receive, insurance payments, negotiating with contractors, and complying with new building codes and inspections. During this process,

Green worried about her regular customers and looked forward to seeing them again. Once the store was open for business, customers started stopping by to visit. As Green reflected, "when I see them [elderly customers], it's a load off my chest...you wonder how they're doing. And when you see them, and to know that they're okay, it's a good feeling. And they're resilient, they're bouncing back."

Green has developed relationships with her customers that go beyond commercial connections. She is not only familiar with the regulars that frequent her store but listens to their troubles and cares about their well-being. This social aspect of the store is reflected in how the customers interact with one another, whether it's chatting as they browse through the clothing racks or sharing information in the checkout line. Green recognized this social aspect of the store and commented, "It, you know, it's unbelievable, but I would say about 75 to 85 percent of my customers, pre-Katrina, are back in some form. If they're not back in the area, they're back checking on property. They'll come in when they're checking on property. A lot of it is social. They come in to socialize, see how everything's doing. They wanna see familiar faces when they come home." Green ensures that the clothing store she manages acts as a social space for women in the Gentilly, Desire, and New Orleans East neighborhoods. The relationships fostered at the store prior to Hurricane Katrina among her customers, and between Green and her customers, built the foundation for social interaction during recovery. By reopening the clothing store, Green facilitated the restoration of these connections.[21] In the language we introduced in chapter 3, the connections made, restored, and fostered at Green's store increased the benefits associated with returning.

Commercial relationships can become overlaid with social content (Granovetter 1985). Storr (2008), for instance, argues that relationships are created and fostered through commercial interactions; that buyers and sellers, as well as coworkers, often become acquaintances, friends, and even spouses. Likewise, Meadowcroft and Pennington (2008) argue that market orders are comprised of impersonal bonds among a large number of people, but that, within this order constituted by thin relationships, thicker bonds can, and do, develop. Additionally, Chamlee-Wright and Storr (2014) apply this argument to the post-disaster context, highlighting how the relationships fostered through commerce can evolve to include mutual assistance, emotional support, and lasting relationships. Green's experiences demonstrate how bonds formed through commerce can withstand disasters, and the possibility that commercial enterprises can serve as social spaces where individuals can reconnect with one another.

Connecting Aid Demanders with Aid Suppliers in Bayswater

Rabbi Mordechai Kruger is a longtime resident of the Bayswater neighborhood in New York City and a rabbi at the Agudas Yisroel of Bayswater synagogue.

The Bayswater neighborhood is a small community, situated on the Rockaway Peninsula in the borough of Queens in New York City. It covers 0.9 square miles and is primarily populated with working-class families. Compared to New York more generally, Bayswater has a lower median income and slightly larger household size.[22]

During Hurricane Sandy, the neighborhood experienced a storm surge from Jamaica Bay, which resulted in up to six feet of flooding. Basements and low-lying buildings suffered major damage, including the Agudas Yisroel of Bayswater synagogue. The neighborhood was without power for two weeks following the storm and conditions were exacerbated further by near-freezing temperatures and the isolated nature of the peninsula. As one Bayswater resident remarked, "On a regular day, it's difficult to get off this . . . peninsula . . . [When something happens], you're not getting out. There is no going back, you're staying."[23] Furthermore, residents who suffered damage also had to deal with the emotional strain of losing family heirlooms and other prized possessions. Another Bayswater resident, Mrs. Russell, recalled how neighbors would console one another in the weeks following the storm. "Somebody's house that was really destroyed down by the water," as Mrs. Russell describes, "her wedding pictures were, like everything she had in her basement was totally destroyed, her wedding album was destroyed, her pictures of her children were destroyed and she just, like walked in and like we did a lot of holding each other and hugging each other, like nurturing each other."

There are eight Orthodox Jewish synagogues in Bayswater, and Rabbi Kruger estimates that there are roughly 700 Jewish families in the neighborhood. While residents of the neighborhood may be members of different synagogues, they share many things in common in addition to their faith, including access to services like Achiezer (discussed in chapter 5), Hatzalah volunteer ambulance service, the Rockaway Citizens Safety Patrol (discussed in chapter 5), and the schools their children attend. They also have Rabbi Kruger in common, who has dedicated his life to the community.

Starting in 2002, Rabbi Kruger founded the Bayswater Neighbors Fund to provide short-term support for those in need. He collects donations, primarily from locals, and then distributes funds to those in need. The small donations can help families purchase food for holidays, pay their monthly bills, or cover a portion of school tuition. While Rabbi Kruger relies on other members of the community to discover when someone needs help and consults the rabbis in the community for guidance, he aims to keep the donations and distributions anonymous. He hopes that the fund fills a gap in the neighborhood, that it helps the working-class families in Bayswater to make ends meet. The people of Bayswater trust him to distribute the funds to families that are honest and are trying to be responsible and to get back on their feet.

Through his work on the Bayswater Neighbors Fund, Rabbi Kruger realized his interest in helping his fellow community members. In addition to continuing to maintain the fund, he has also worked as a caseworker for

Metropolitan Council on Jewish Poverty (Met Council) and is now a director of a nonprofit that helps Jewish adults identify their career goals and gain the training and experience needed to fulfill their goals.

In the aftermath of Hurricane Sandy, Rabbi Kruger utilized his experience and skills to help the Bayswater community recover. He became the representative for Bayswater for the Community Assistance Fund (CAF), which Rabbi Bender and Achiezer organized (discussed in chapter 5).[24] He recalls how he assumed the position: "36 hours [after the storm, Achiezer was asking] volunteers to coordinate distributions of funds, and I don't know if anybody told me to go, I just kind of showed up... [and said] okay, I will do it." The trust he had built in the community helped him reach out to those who needed help. By utilizing trusted members of the community, including rabbis and other community activists, Rabbi Kruger was able to spread the word about CAF, encourage applications, and help the community receive the supplies and resources it needed to recover.

Rabbi Kruger worked with a couple in the neighborhood to assess needs and document them in a spreadsheet that could be used to request and direct donations. For instance, if someone wanted to donate mattresses, Rabbi Kruger could look at the spreadsheet to see how many mattresses were destroyed in the storm. Explaining the benefits of the spreadsheet, Rabbi Kruger stated, "What that ended up doing was that there were fund givers who would come forward with specific interests... [W]e were able to pull that information... we can tell you how many of those we need... [T]hat made it a lot easier to approach funders and it really sped up the relief effort."

Additionally, Rabbi Kruger relied on the volunteers at Young Israel of Wavecrest and Bayswater, a synagogue in the neighborhood that became a relief center after the storm, to talk to residents and encourage them to apply for resources. They helped spread the word about CAF by sending emails, handing out fliers, and checking on the elderly by going door-to-door. One person that Rabbi Kruger relied on was fellow resident of Bayswater and longtime friend, Mrs. Russell, who ran the kitchen at Young Israel after the storm (her efforts will be described in the next section).

Rabbi Kruger also took on fund-raising efforts to repair Agudas Yisroel of Bayswater, which had sustained severe damage from the storm. The synagogue building has its main floor at basement level, with steps from the street leading down to the main floor entrance. This design made it particularly vulnerable to flooding. During Hurricane Sandy, the floodwaters surged into the building, collapsing the entrance and destroying holy books. The congregation was able to raise funds to repair the building and replace important texts, and now hopes to build a second story to be able to withstand future storms.[25]

The tight-knit community of Bayswater had strong, preexisting ties prior to Hurricane Sandy. After the storm, Rabbi Kruger was able to use his experience and position in the community to assess damage, help funnel resources to the community, and encourage a quick recovery. He, with the help of his neighbors, spread the word about CAF and other resources by sending emails, handing out fliers, and talking to residents at the relief center at Young

Israel and going door-to-door. They tracked needs on a spreadsheet, distributed funds, and donated resources. Through these actions, Rabbi Kruger sustained and restored the social network of the Bayswater community and connected aid suppliers with aid demanders. In the language of chapter 3, Rabbi Kruger's efforts significantly lowered the costs of rebuilding for disaster victims and increased the probability of recovery by supporting and reinforcing the ties that make Bayswater a tight-knit and unique community.

If communities are to recover, it is important that aid demanders be linked with aid suppliers. Sobel and Leeson (2007: 520), for instance, examine the market for disaster relief and argue that "Relief demanders know when relief is needed, what they need, and in what quantities, but they do not necessarily know who has the relief supplies they require or how to obtain them. Similarly, relief suppliers know what relief supplies they have and how they can help, but they may be largely unaware of whether relief is required and, if it is, what is needed, by whom, and in what locations and quantities." They argue that effective relief management must generate appropriate levels of knowledge at each of three stages—the recognition stage, the needs assessment and allocation stage, and the feedback and evaluation stage—and that only decentralized, private decision-making can effectively generate the appropriate knowledge (ibid.). Likewise, as Bolin and Stanford (1998: 26) note, "what people need after a disaster is a complex issue and cannot, in any sense be answered unambiguously... Unmet needs, in a general sense, can trouble people's day-to-day lives, and a disaster generally compounds those needs and troubles." In their examination of the 1994 earthquake in Northridge, California, Bolin and Stanford (ibid.) found that NGOs and community-based organizations were able to utilize "local knowledge and expertise, and establishing progressive working relationships with city, county and federal agencies, were able to implement programmes that connected victims with unmet needs to a diversity of resources" (ibid.: 34). Rabbi Kruger's actions further exemplify the importance of local knowledge in connecting aid demanders with suppliers and spurring community rebound.

Creating a Social Space through Emergency Relief in Bayswater

Shaindle Russell, or Mrs. Russell, as she is referred to by her neighbors, is a longtime resident of Bayswater. She has lived in the area for over 20 years along with her husband and eight children. Her commitment to the community is exemplified through the jobs she has held. She has worked as a grants administrator for a Jewish day camp, managed a prekindergarten program at a local Jewish girl's school, and is currently a doula and birth coach. In many ways, Mrs. Russell is a caregiver, and she performed that role for the Bayswater community after Hurricane Sandy.

Just before to the storm, Mrs. Russell had gone to Manhattan with some of her children. They caught the last train home, just as the weather started getting worse. Once back in Bayswater, they gathered flashlights and water and prepared to sit out the storm. They heard a loud explosion, and the electricity went out. Her son, a member of the Rockaway Citizens Safety Patrol (RCSP) at the time, and her husband went out to check on others in the neighborhood and, as the floodwaters started to rise, encouraged residents to wait on the second floor of their homes.

The next morning, Mrs. Russell realized they did not sustain any flooding and went for a walk with a friend to check out the rest of the neighborhood. They came upon Agudas Yisroel of Bayswater and mourned the devastation of the synagogue and holy books. It was then that she realized the damage sustained by her neighborhood. In a piece for *Jewish Action*, Mrs. Russell (2013) recalled, "That's when it hit me: my house was fine, my family was fine, but my neighborhood wasn't. I had to help."

They then went to Young Israel of Wavecrest and Bayswater. Mrs. Russell remembered observing men taking food from the synagogue, "So there was a delivery of milk left for the boys...and bagels. And I saw people, men walking out of the prayer service just taking milk and bagels. There were no boys here to take it. So I was like, in my mind, we are going to have a food issue here because there is no electricity and from what I heard, the whole island, the whole peninsula here was out of electricity." And, Mrs. Russell was right. Recall that her neighborhood and much of the Rockaway Peninsula was without power for two weeks.

She then went inside the synagogue and spoke to the rabbi at Young Israel about addressing the need to provide electricity and food to residents. As Mrs. Russell recalls, "I said, 'You are going to have a food issue.' So he goes, 'Okay, we are opening a food pantry and you are in-charge.' So I was like, 'Okay, no problem.'" Mrs. Russell got right to work preparing the kitchen and calling in requests for donations. She explained that within a few hours they received food from the Jewish Community Council, and by the third or fourth day they were serving 300–400 people three hot meals a day as well as snacks. She received food from catering companies in Brooklyn and could place orders for fresh produce and other goods from a local grocery store. Mrs. Russell and three of her friends ran the kitchen from 7 a.m. to midnight every day for two weeks. They set up the food, cleared dishes, and talked with residents. In all, they supplied hot meals and snacks to neighbors and became a distribution center where residents could pick up nonperishable food and other resources.

Mrs. Russell also recognized that Young Israel had an important social function to perform after Sandy beyond its disaster relief efforts. As Mrs. Russell noted, "So people would come in between meals and they wanted cake and coffee and fruit and it was a really nice place for people to hang out." With the help of a generator, the building had power to keep lights on into the evening, and many residents gathered there to talk about their day, plan for the future, and forget about some of their immediate troubles.

Rabbi Kruger, of Agudas Yisroel of Bayswater and the Bayswater representative for CAF, recognized that Young Israel had become an important space in the community and asked Mrs. Russell to help him spread the word about CAF. As Mrs. Russell recalled, "[Rabbi Kruger] saw that I really had that in control so then he goes, 'Do me a favor, I have these applications. Can you work the crowd and find out who needs money and whose houses were destroyed and we can get them the money.'" So after the food was served and people were eating and socializing, Mrs. Russell would walk around and talk to residents about CAF. "I was able to work the crowd and make sure that people filled out applications and I had to convince some people because they didn't want to take from anybody," she explained.

Mrs. Russell was dedicated to her community after the storm, working 17 hours a day, and only taking breaks to bring food home and check on her family. Rabbi Kruger remarked, "Mrs. Russell has been a powerhouse, just all kinds of help for who knows who, for years and years. I'm a big fan of Mrs. Russell." Rabbi Kruger also refers to Mrs. Russell as having an "antenna." Although we do not contend that she has superhuman powers, Rabbi Kruger's remark is a reflection of Mrs. Russell's dedication to the community. In fact, he frequently relies on her connections to the community to determine which families need help and should receive funds from the Bayswater Neighbors Fund, and he did the same when spreading the word about CAF after Hurricane Sandy.

Further, her children helped too. Her son volunteered for the RCSP and her daughter helped answer phones at Young Israel, which had turned into a command center for the patrol, a resource distribution center for donations and supplies, as well as a place to get a hot meal, use electricity, and share information. Their hard work did not go unacknowledged, and people reciprocated by giving them a generator and gasoline, so they could heat their house. Mrs. Russell also has a foster child that has cerebral palsy, and the generator kept his nebulizer, feeding tube, and chest compression machine running (Russell 2013).

By providing hot meals and companionship to her neighbors, Mrs. Russell helped them deal with their loss, kept up their spirits, and encouraged them to apply for funding through CAF. She worked the kitchen at Young Israel until the power came back in Bayswater, roughly two weeks after the storm hit. By that time, CAF resources were beginning to be distributed and residents could access food, supplies, and other resources on their own. In the language that we introduced in chapter 3, Mrs. Russell's efforts lowered the costs of recovery by not only providing needed goods and services but by also fostering a social space immediately after the storm and encouraging residents to apply for assistance.

Similarly to Cicek opening a coffee shop in Chalmette after Hurricane Katrina, Mrs. Russell and others in Bayswater not only provided nourishment but also a social space where residents could socialize after Hurricane Sandy. While the food pantry and kitchen were not a commercial enterprise, it filled an immediate need for shelter, food, and electricity and also provided a space for residents to reconnect, receive information about assistance, and formulate plans for recovery. Mrs. Russell was crucial to reconnecting the social networks in Bayswater.

In Summary, Entrepreneurs Restore and Create Social Networks after Disasters

As the stories above highlight, entrepreneurs play a crucial role in restoring social networks after disasters. Whether bringing together close-knit homogeneous communities (restoring bonding social capital) or providing connections across heterogeneous individuals and groups (repurposing bridging and linking social capital), entrepreneurs find ways to reach out to dispersed residents and encourage coordinated recovery.

Throughout New Orleans, entrepreneurs were utilizing preexisting social networks to exchange information and stimulate recovery. LaToya Cantrell, as the president of the Broadmoor Improvement Association, reached out to displaced residents and initiated rebound to fight the plans to turn Broadmoor into a green space. Not only did Broadmoor prove viability, it also received funding and support to bring additional services and infrastructure to the community, such as a new charter school and a renovated library. Similarly, Ben Cicek and Sandra Green restored social networks in St. Bernard Parish and Gentilly, respectively, by establishing businesses that served as social spaces for their communities. Cicek opened a coffee shop as a way to provide food, coffee, and Internet access for locals. Not only did he succeed in his business venture but he also established a space where residents and workers could discuss their problems and exchange information and resources about rebuilding. Similarly, Green reopened a women's clothing store that also served as a social location for her regular customers.

Furthermore, in Far Rockaway, New York, Rabbi Kruger and Mrs. Russell facilitated the exchange of information and helped coordinate recovery after Hurricane Sandy. Rabbi Kruger worked as a representative for the Community Assistance Fund and developed ways to assess damage and request supplies. Mrs. Russell, while working to provide meals for the neighborhood, also spent time talking to residents about their problems and reconnected them with aid suppliers.

Each entrepreneur in this chapter went about helping community members restore social networks in different ways. Yet, each spurred recovery by creating opportunities for community members to spend time together and by connecting community members with individuals who might provide information and assistance.

Chapter 7

Entrepreneurs as Signals of Healthy Community Rebound

As discussed in chapter 3, disaster victims will want to know if others plan to return when deciding whether to return and rebuild or to relocate somewhere else. If everyone is waiting for signs from others before deciding how to react, recovery might never occur (Chamlee-Wright 2010; Chamlee-Wright and Storr 2009a, 2010a). Entrepreneurs, however, can act as "focal points" or "points of orientation" for residents as they decide whether or not to return and rebuild after a disaster and formulate their post-disaster recovery plans.

Entrepreneurs can signal to displaced residents that community rebound is likely to occur, and they can actively encourage others to return. They can do this in a variety of ways. Entrepreneurs can act as first movers, accepting the risks associated with returning before the recovery process is under way. They can also lay the foundation for return by opening enterprises that provide the goods, services, and opportunities that have to be in place before some displaced residents could seriously contemplate returning. Additionally, entrepreneurs can facilitate communication among displaced residents and organize recovery efforts. Moreover, they can simply reopen their enterprises and, in so doing, signal a return to normalcy.[1] Returning to the language used in chapter 3, entrepreneurs can change the calculus for residents attempting to overcome the collective action problem associated with recovery. By connecting residents and obtaining resources for the community, they reduce the costs of returning and rebuilding. By providing support and resources, they increase the benefits of returning. And, by emphasizing the important and cherished aspects of their community, they reinforce the belief that others will also return.

Importantly, the Hurricane Katrina victims we interviewed described (1) the nature of the collective action problem that followed Katrina as well as (2) how key figures helped them to coordinate on return. The Hurricane Sandy victims that we interviewed also highlighted the important role certain individuals in the community played in helping to spur disaster response and recovery.

Scholars have drawn attention to the coordination challenge that must be overcome if disaster recovery is to occur, as well as the important role commercial, civic, and community leaders can play in helping communities coordinate their recovery efforts. Much of the literature suggests that homogenous, tight-knit communities will have an advantage over more diverse, less-connected communities in solving this collective action problem and bringing about community rebound and redevelopment (Aldrich 2011a, 2011b; Chamlee-Wright and Storr 2009a, 2010a; Murphy 2007). Chamlee-Wright and Storr (2009a, 2010a), for instance, examine how the Vietnamese community in New Orleans East was able to overcome the collective action problem and return and rebuild quickly after Hurricane Katrina, in large part, due to the persistence and dedication of Father Vien, the pastor of the Mary Queen of Vietnam Catholic Church. However, other research has highlighted how more diverse and loosely connected communities can also overcome the coordination problems that hinder rebound (Bolin and Stanford 1998; Chamlee-Wright and Storr 2010a; Murphy 2007; Shaw and Goda 2004; Storr and Haeffele-Balch 2012). Storr and Haeffele-Balch (2012), for instance, examine the recovery in Broadmoor after Hurricane Katrina and found that the leadership of LaToya Cantrell and the preexisting organization of the Broadmoor Improvement Association allowed residents to communicate with one another and collectively prove the viability of their neighborhood to post-disaster planning authorities. Commercial, civic, and community leaders and community-based organizations play an important role in encouraging recovery and addressing the specific needs of affected communities after disasters.

Leading Return and Recovery in Village de l'Est

Father Vien is the pastor of the Mary Queen of Vietnam (MQVN) Catholic Church in the Village de l'Est neighborhood of New Orleans East. Before Hurricane Katrina, MQVN had 3,800 Vietnamese-American parishioners, many living within a mile of the church.[2] There were also 75 businesses in the community, most of which were owned by parishioners. This tight-knit and insular community was formed by migrants from Vung Tao, a fishing village in South Vietnam, who sought a new home in the United States after the fall of Saigon in 1975.[3] The church, built in 1985, is the focal point of the community. In addition to holding mass twice a day, the church also holds Vietnamese classes, study groups, luncheons, and other activities. As Chamlee-Wright and Storr (2010a: 155–156) argue, "MQVN is at the spiritual, social, commercial and geographic heart of the New Orleans East Vietnamese community." At the time of Katrina, Father Vien was the uncontested leader and spokesperson for the community, despite having no official political position.

Prior to Hurricane Katrina, the MQVN church regularly hosted Vietnamese cultural events and weekly farmers markets. Though median incomes still remained below that of Orleans Parish, home ownership rates were nearly 40 percent in 2000.[4] By 2005, the church was taking the lead in advancing an ambitious set of community redevelopment projects, including Viet-Town (a Vietnamese cultural and commercial center), revitalization of the two Vietnamese shopping centers in the community, establishment of an urban gardening project, and building of a housing development for the elderly adjacent to the church.

Father Vien became the pastor of MQVN parish in 2003, and was supported by two other priests and five paid staff members. Like MQVN pastors before him, his leadership not only stems from the religious authority he holds, but also from his personal charisma and willingness to take on the role as spokesperson for the community (Chamlee-Wright and Storr 2009a). His commitment to the community is exemplified by his efforts during and after Hurricane Katrina.

As previously mentioned in chapter 3, Father Vien chose to weather the storm with 500 of his parishioners who did not evacuate before the storm by taking shelter at the church. Once the storm passed, residents started returning to their homes to check for damage. But a few hours after they returned home, the neighborhood began to flood due to the levee breaches. With the floodwaters rising rapidly, Father Vien and several volunteers contacted residents on their cell phones and used boats to pick them up and take them back to the church, where they took shelter in the second floor of the school building. Then, in the days and weeks following the storm, Father Vien visited evacuation sites in Arkansas, Georgia, Texas, and Louisiana to check on his parishioners, share information, and encourage them to return home.

The neighborhood sustained significant damage from flooding. According to the Urban Land Institute (ULI), the principal consulting group advising the Bring New Orleans Back (BNOB) Commission, "The New Orleans East area experienced some of the city's most severe flooding, with flood depths ranging from five to more than twelve feet" (ULI 2005: 43). Based on such reports, the BNOB Commission recommended converting the area into open space, connecting it to the Bayou Sauvage National Wildlife Refuge (ibid.). However, by the time the BNOB Commission made their recommendation, the MQVN community was already returning and on the way to recovery.

Father Vien worked to encourage a collective sense among his parishioners that the MQVN community was a "second homeland" and that the Katrina experience was linked to their migration experience out of Vietnam. Father Vien also made his parishioners believe that the community would rebound after Katrina. Despite the extensive damage, within six weeks after Katrina, Father Vien began holding church services at MQVN for parishioners who had already returned, and issued calls for those still displaced to return home. Through these actions, Father Vien sent a powerful message that the MQVN community would recover from the storm. At a time when many parts of New Orleans were essentially deserted, these services attracted 300 attendees on October 9 and 2,000 attendees by the end of October. And, parishioners often cited Father Vien as the impetus for their return. For example, Chinh Do† reflected, "Father Vien, he's why we're staying. He put this community back together. He's a powerful man, too. The way he sounds, it's like he knows. We believe in him and we trust him too. What he says, we do it." As Chamlee-Wright (2010: 62) explains, "many people reported that they came back because Father Vien told them to." And, "deference to Father Vien's calls for rebuilding directly impacted people's decisions. In turn, because people were confident that others would heed the pastor's call to return, they felt more assured that they would not be the only ones to return" (ibid.: 63).

Father Vien and the other priests and church staff utilized the strength of the MQVN community to not only encourage parishioners to return but also to signal to government officials that they would not let their neighborhood be turned into open space. They were interviewed by the media who photographed the assembled parishioners attending Mass just weeks after the storm. They made it known that they were not going to wait for government approval to rebuild, and that the community was capable of recovering on its own. Residents echoed this sentiment. For instance, Thau Vu† stated, "We didn't wait for the city to decide what [was] going to happen to our community. We decided to take it in our own hands of what will happen... We didn't have to wait for FEMA to give us money... And, we didn't rely on the city to tell us [if we could] come back;... [if] we can start doing repairs... We just came back, did repairs and you know, got on with our lives."

Father Vien, the assistant priests, and the church's lay leadership also played a major role in providing food, supplies, and support after the storm. After the government and nonprofit emergency relief organizations failed to provide assistance that addressed the cultural and linguistic needs of the community, they established the MQVN Community Development Corporation (MQVN CDC).[5] At first, the MQVN CDC took over case management for Small Business Administration loans, Road Home assistance programs, and other disaster assistance from Catholic Charities. Then they focused on securing funding for larger community initiatives, such as a retirement center, an intercultural charter school, an outpatient medical clinic, the reopening of a nearby hospital facility, and the development of a business corridor. Throughout this process, the MQVN CDC staff became adept at grant writing and fund-raising, securing over two million dollars in capital to rebuild and expand community businesses.

The MQVN CDC also made sure to inform the BNOB Commission of the developments and represented the community at planning committee meetings. By inserting themselves into the city planning process, the MQVN CDC ensured that the MQVN community would be included in future plans instead of proposals to turn the area into an open space.

These efforts allowed Father Vien to focus his attention on the political and logistical issues that hampered recovery. He overcame numerous hurdles in the months following Katrina, which included (1) proving the need for electricity to be returned to the neighborhood, (2) obtaining FEMA trailers and city approval to open a trailer park for residents while their homes were repaired, and (3) preventing a landfill from opening next to the community.

First, when city officials rebuffed requests from Father Vien to restore electrical service to the community, he went directly to the local power company,

Entergy. His ability to organize the community enabled him to convince the company of their community's need for power and get the electricity back quickly. As Father Vien explains:

> [I]n order to justify [and] divert power out here, we must justify that there are people here planning to receive it...[T]hey needed paying customers...I gave him pictures that we took of our people in Mass, first Mass. First Mass was 300 [people], second Mass was 800, third Mass we invited all the people from New Orleans, and we had more than 2,000...He said, "Those I get. But now we need a list." And so we went and got what he asked...So within one week, I went back to Lafayette, we went back to his office, I said, "Well, the city has 500 petitioners."

The community got electricity restored by the first week of November 2005, when other communities in Orleans Parish did not yet have power.

Second, Father Vien helped to facilitate and receive FEMA trailers for the community. He received permission from the archdiocese to use an area of land that had been slated for a senior housing project. While FEMA approved the request fairly quickly, gaining the necessary permit from the city proved more difficult. Father Vien used his professional connections and political know-how to get resources for the community. As Father Vien explains:

> [We acquired permission from FEMA] on the 19th of October. We got the legal [documentation], and then we did the paperwork and brought it to the mayor's office...The mayor refused to sign it...[T]hey had to bring it up at the [Bring New Orleans Back] commission's meeting on...Monday [November 21st]. And so I called the archbishop, because he was on the commission. So I called the archbishop and told him that unless the mayor signed it on that day, we will set up a tent city because my people are living in moldy homes waiting for that. And so...that evening, the archbishop called me and said, he said that he did it. [The mayor] signed it. [But FEMA never received the signed documentation from the mayor's office.]...So I called the archbishop and asked him to contact the mayor and have the mayor fax it to his office...But nothing moved for a whole week...finally when I realized what they were doing, I called them again and I said that Monday, "if it doesn't happen, that [tent] city's going up."

Father Vien also spoke up at City Council meetings, questioning whether the mayor had a racial bias, contrasting the mayor's pleas to African Americans to return while refusing to allow FEMA to provide temporary housing to the Vietnamese Americans who had already returned. These charges captured the attention of several City Council members as well as then US Senators Hillary Clinton and Mary Landrieu, who were in Louisiana at the time. The political pressure to prevent a tent city before winter arrived and to stem accusations of racial bias forced the mayor to grant the permit to establish a FEMA trailer park for the residents of the MQVN community.

Third, shortly after the FEMA trailer park was approved and opened, the city announced plans to open a landfill in the area adjacent to the MQVN community. Since they faced a serious issue with debris removal from the storm and would receive 22 percent of the revenues from landfill if it were to be located within the city limits, the city had an incentive to develop the landfill.[6] The landfill, however, posed an environmental hazard and threatened development plans for the MQVN community. In response, Father Vien urged the youth of the community to form the Vietnamese-American Youth Leadership Association (VAYLA) to address the initiative through political activism. One young resident, Ngoc Hoang† recalled, "[Father Vien] has a strong will...that determination, passion for the youth to take charge. He encouraged us to enter politics and be assertive. Me, I just want to live my own life...it's a Vietnamese characteristic; we want to live in peace. Politics [is] stress and lies, [but] Father Vien has asked us to take the lead...if you don't speak out, if you're not united [as a community] people don't respect your rights." VAYLA, comprised of young residents with bilingual skills, was able to coordinate demonstration efforts and connect with local and national environmental and community rights organizations, including the Tulane Environmental Law Center, the Sierra Club, the Louisiana Environmental Action Network, and the Louisiana Justice Institute. For instance, they organized rallies at City Hall and barred dump trucks from entering the landfill site. The community also joined other organizations in petitioning the United States Federal Courts, the New Orleans City Council, and the Louisiana State Senate. Furthermore, Father Vien testified before the US Senate Committee on the Environment and Public Works voicing concerns over environmental threats of the landfill to the community (Vien 2007). These efforts lead to the eventual defeat of the landfill initiative.

It is clear that Father Vien acted as a focal point for community rebound and spokesperson for the MQVN community after Hurricane Katrina. In fact, virtually every member of the MQVN community that we interviewed reported that the local church leadership was important to the community's rebuilding success. They pointed to both their religious authority and the distinct signaling role they played in getting residents to return. Father Vien's leadership skills were noted outside of the community as well, media described him as a priest who "personifies resilience" (Pope 2006) and an "instrument in helping people reclaim their lives" (NPR 2011).

Had Father Vien not acted, the dynamic in the MQVN community would likely have been very different. Fortunately, Father Vien did act and spurred a rapid recovery for the MQVN community. Less than two years after Katrina, the overwhelming majority of his parishioners had returned, most of the businesses they owned had reopened, and the community was well on its way to being rebuilt. In fact, by the summer of 2007, about 90 percent of the residents had returned to the MQVN community while the repopulation rate in New Orleans overall was only 47 percent. By expanding his role from spiritual leader to community spokesperson and political activist, Father Vien ensured the MQVN community returned and was taken seriously by the political actors in charge of the overall recovery of New Orleans.[7] In the language we deployed in chapter 3, by rallying residents and emphasizing the importance of the MQVN community, Father Vien significantly reduced the costs of returning and increased the benefits of returning as well as the probability that others would return.

Research on the MQVN community and the efforts of Father Vien reinforces the notion that community leadership and "build it and they will come" strategies that he adopted helped solve the collective action problem (Chamlee-Wright 2007, 2010). For instance, Chamlee-Wright and Storr (2009a) examine how the tight-knit nature of the MQVN community fostered the development of club goods that fit the unique needs of the community, and how these club goods were crucial to accessing resources and overcoming the burdens of rebuilding.[8] Similarly, Chamlee-Wright (2010) highlights how the collective narratives and experiences of the MQVN community, including the strong ties to the church and viewing the community as a second homeland, enabled recovery. And, Aldrich (2012) highlights how Father Vien and other local activists maintained and fostered the bonding social capital needed for a speedy recovery.

Rebuilding Schools as a Precursor to Recovery in St. Bernard Parish

Doris Voitier, superintendent of the St. Bernard Parish Unified School District, recognized that getting the schools back up and running as quickly as possible after Katrina would be crucial to having families and businesses return to the area. In fact, in the middle of October, despite the fact that most of the district's 15 facilities were severely damaged, she pledged that there would be a space for every student that signed up for classes at the November 1, 2005 registration.

Recall that St. Bernard Parish is a predominately white, family-oriented community, filled with hardworking individuals (Chamlee-Wright and Storr 2010a).[9] As Voitier describes, "St. Bernard Parish is a community of about 68,000 people prior to the storm. Our public school system . . . it was basically a blue-collar hard working community. Fewer than I think 10 percent have college degrees actually."

In preparation for the storm, the parish had set up a shelter at Chalmette High School, and about 250 people came to ride it out. Voitier explained that the high school was prepared to be a shelter; "they came in, we registered them; we're pros at this." Once the storm had passed, Voitier recalled that they believed everything was going to be okay. Then one of the firefighters at the shelter heard over his radio that the levees had been breached. Within minutes a wall of water was rushing down the street. They got everyone upstairs to the second floor but ran out of time to secure most of the provisions.

The very next day, Voitier went to the State Department of Education and met with the State Superintendent of Education. He gave her space in the office to set up shop, and she got straight to work. Her first priority was to make sure her employees continued receiving their paychecks, so they could return and rebuild their homes. Her second priority was to start compiling records and assessing damage to get the schools up and running again. As she explained, "I had many things on my mind at that point: all of our employees, all of our teachers, all of our people, I knew were scattered throughout the country. And, I knew they had lost all their homes, had nothing. And, I wanted to make sure that we could pay them as long as we could pay them…I knew our students were scattered all over the country…So there were some things that we had to take care of immediately before I could even come back and assess damages."

When Voitier was finally able to assess the damage, she realized that the level of damage her community suffered was immense. Recall that St. Bernard Parish had up to ten feet of floodwater, and the parish had been underwater for weeks after the storm, resulting in over $3 billion in property damage. As a pastor and longtime resident of the area exclaimed:

> It was beyond belief, and just in St. Bernard Parish alone you're dealing with 42 square miles of destruction. You're dealing with 26,900 homes, 3,000 businesses, four major high schools, and you're dealing with five out of six people that did not have insurance, because they were in federally declared non-flood zones. I personally bought—me and my wife owned a home for 18 and a half years, a couple of blocks from here in a flood zone, and paid flood insurance all our life. Nine months before the storm, we bought our dream home in a non-flood zone and lost everything we owned. And, God has been very gracious to us, and as a pastor, many people come to your aid, but I know that a lot of people have suffered because of this situation.[10]

When Voitier initially pledged to open the school for any students that registered by November 1, she expected maybe 50 students would actually register. As she told *ABC News* (2006) in a profile on her efforts, "There was not a home, a church, a school or a business that was habitable. Every one of our 15 school sites was destroyed."

Despite the damage, 703 students registered and committed to returning to school sometime between November 14, when classes would resume, and the start of the new semester in January. In the words of Voitier, "They wanted to be back here." By the end of January, over 1,500 students had returned, and by the end of the school year, 2,360 students; roughly one quarter of the pre-Katrina enrollment of 8,800 students were back in school. As Chamlee-Wright and Storr (2010a: 156) posit, "By committing to reopen the schools on a particular date, Voitier signaled to displaced residents that the community recovery would, indeed, occur and allowed residents to make clear decisions about whether and when to return."

Voitier, however, had to overcome numerous hurdles and setbacks in order to fulfill her promise of having space available for the students that returned. At first, she attended numerous meetings and tried to coordinate with FEMA and other authorities. As an administrator of Chalmette High School recalled, "[Voitier] was going to a gazillion different kinds meetings, you know the State Department of Education, FEMA people…And the answers she was getting just didn't sit well in terms of what we knew we had to do and so ultimately she said, 'you know, never mind, we'll just do it ourselves and I'll send them a bill, eventually.'"[11] Similarly, Voitier recalls that she finally said, "The heck with you guys. We can do it, we'll make it happen, and we'll send you the bill." But then she realized they had a lot of work ahead of them, and recalls, "I just sort of sat back and said, 'Okay. That was very cavalier. You know, how are we gonna' do this?' So I got a local contractor."

Voitier, working with a local contractor, used funds from the school district to clean out debris and purchase resources and supplies. They ordered portable classrooms to be delivered from Georgia and Carolina. With generators to provide lighting, they set up the portable classrooms in the parking lot of the high school. When more children showed up than could be housed in the portable classrooms, they used the second floor of the high school while the first floor was being renovated. In addition to providing space for classes, Voitier also had to encourage faculty and staff to return. So she bought trailers for them to live in while they rebuilt their homes.

Within a year and a half, Chalmette High School was completely redone and a second high school, Andrew Jackson High School, was rebuilt and was used to house elementary school students. At that time, about 1,750 middle school and high school students attended Chalmette and around 2,000 elementary school students went to Andrew Jackson. They also found ways to provide medical services for the children by partnering with doctors from the Louisiana State University School of Health and Allied Science as well as Tulane University.

While the schools would have never reopened if Voitier had worked on her own, her ability to inspire, lead, and mobilize resources was crucial to the recovery of St. Bernard Parish. As another administrator from Chamlette High School noted, "[T]here is no question... Doris' leadership was primary."[12] Furthermore, Voitier has been recognized by the media and received the 2007 John F. Kennedy Profile in Courage Award™ for "her courageous fight to rebuild the schools of St. Bernard Parish after Hurricane Katrina, despite pervasive devastation and bureaucratic indifference" (*ABC News* 2006; *Source for Learning News* 2007).

Voitier recognized that she could encourage St. Bernard Parish residents to return by ensuring that their children could attend school. In the language we utilized in chapter 3, Voitier increased the probability that displaced residents would return by providing a needed service that also served as a strong signal that recovery was under way. Furthermore, she would not be deterred when faced with setbacks and delays from government authorities. Rather, she realized that she could rely on herself, her employees, and the community to do what needed to be done.

The restoration of public services is important in reducing the uncertainty after a disaster. As mentioned in chapter 3, it is reasonable for residents to wait until services, such as utilities and schools, have been restored before making the decision to return. However, it is also reasonable for city officials to wait until people have signaled that they will return before investing in restoring such services. In that scenario, an entrepreneur who is willing to commit to ensuring that needed services are available can help communities overcome the post-disaster collective-action problem and encourage community rebound. Voitier did just that by committing to reopen the school. Chamlee-Wright and Storr (2010a) highlighted Voitier's efforts as an essential example of how social entrepreneurs can encourage recovery. And, Chamlee-Wright and Storr (2011b) examined how Voitier stuck to her commitment to reopen schools, even if that meant she had to rely on herself and her own staff instead of using official disaster relief channels.

Signaling a Return to Normalcy in the Lower Ninth Ward

Casey Kasim is a longtime resident of and business owner in New Orleans. He and his partners have opened numerous gas stations, convenience stores, and laundromats in the greater New Orleans area. Kasim, his wife, and six children have made New Orleans their home and established themselves in the community through their business efforts.[13] After Katrina, Kasim's efforts to reopen his business at a time when it was unclear whether or not the Lower Ninth Ward would actually recover served as an important focal point for community rebound.

In 2003, Kasim decided to invest in a business venture in the Lower Ninth Ward and opened a gas station and convenience store, called Discount Zone, on the corner of St. Claude Avenue and Reynes Street. As we described earlier, the Lower Ninth Ward is a close-knit, fairly homogeneous community. As Alice Craft-Kerney (whose post-disaster efforts were discussed in chapter 5) summarized, "Truly the Lower Ninth Ward was a very family oriented place. We had our problems because it was part of the urban setting. But, this was the more country part, if you want to consider [it part] of New Orleans. We have the highest rate of home ownership in the entire city." Recall that the Lower Ninth Ward had a pre-Katrina population of 14,000, was 98 percent African American, and also had a home ownership rate of 60 percent. But 36 percent of the population lived below the poverty line, and the community suffered from high crime rates.

Kasim noticed a lack of amenities in the area and attempted to fill that gap by opening a one-stop shop with gas, laundry facilities, groceries, take-out food, and other items. He saw the opportunity to provide needed amenities to a community that lacked them and invested two million dollars into developing the facility and ensuring it was clean and secure. He also chose not to sell alcohol to reduce loitering. Looking back at the decision, Kasim explained, "[T]he impression of this neighborhood was very bad, very rough, that only the criminals lives here. But it was just the opposite... I would even say 95 percent were very good people. A lot of people told me, 'You're crazy. Why did you invest $2 million here?' I told them I have lived with... African-Americans. They are good people. You do not need to be scared because they are not the same color like you." The venture was a success and within a few years the Discount Zone became an important social space in the community of the Lower Ninth Ward (Chamlee-Wright and Storr 2008, 2014). As Kasim describes, "[Most of my customers] were locals that come every day and you see them and you get to know them by name. You know their problems, you know their children, you socialize with them."[14]

As Hurricane Katrina approached, Kasim and his wife sent their children to Texas with family members but elected to stay behind and protect their Lower Ninth Ward property. Kasim was downstairs in the convenience store when the levees broke, and the floodwaters rushed into the community and into his store. He weathered the storm and the flooding inside his boat, which he had the forethought to tether to his store. Once the conditions calmed down, he climbed to the second floor apartment to his wife. According to a profile in *American Coin-op*, "they immediately started hearing neighbors screaming for help and Kasim jumped back in his boat" (Nagle 2008). As mentioned in chapter 3, Kasim used his boat to rescue over 30 people who were stranded in their homes. He brought them back to Discount Zone, where they took shelter in the second floor and salvaged food and water from the store.

The Lower Ninth Ward suffered severe damage compared to other areas in New Orleans after Katrina. Multiple levee breaches brought a 20-foot surge of water into the community and led to extensive flooding of up to ten feet throughout the neighborhood. The surge pushed out homes from their foundations and flattened entire neighborhood blocks. According to damage assessment surveys conducted in June 2006, over 3,000 of the 4,750 residential buildings in the area were beyond repair.[15] As one resident recalled, "We went first looking at the house, the city. Just looking at it. And it looked like war zone. That's what it looked like."[16] As another resident described, "I lost everything in the house. I lost all my cars, my truck, and that motor home you see sitting over there."[17]

The combined level of destruction and the persistence and concentration of poverty in the Lower Ninth Ward led many—including government officials—to conclude that the neighborhood would not rebuild. Chamlee-Wright and Storr (2009b: 616) observed, "While most Gulf Coast residents were allowed access to their property within weeks of the storm, residents in the Lower Ninth Ward were not permitted access for 3 months, when the contents of the homes continued to fester causing further damage." By prohibiting residents from returning, delaying the restoration of public services, and doubting whether the community would be able to rebuild at all, officials only added to the "signal noise" associated with recovery (Chamlee-Wright 2007).[18]

The community, however, felt differently, and many wanted to return. Again, Kasim saw an opportunity to fill a gap and reopened his store to provide the amenities that others needed to return and rebuild. As Kasim explained, "[I]f there is a gas station or convenience store in your place, you feel that there is life in it, if there is none—it would look... dead. If there is someone who can provide [gas and groceries], they can live around [there] and they can go on with their lives."

Several months after the storm, Kasim opened a smaller facility a few blocks away from Discount Zone. Then in the spring of 2007, he reopened the main facility. Kasim noted that the process was long and filled with uncertainty, including applying for a loan, getting turned down, and then ultimately getting a loan from the Small Business Administration. Yet, throughout the process, his customers and the community supported and encouraged his efforts. He describes how customers came to the store and said, "[P]lease, we will give you as much of the business, support you and stay with you, but just be patient and things will be back because we need your service, we need your gas," and "if you are staying, you are going to encourage other businesses to come back."

Kasim served as an important focal point for recovery in the Lower Ninth Ward. Prior to Katrina, Kasim had established himself in the community by providing amenities that had previously been lacking, such as gas, food, and laundry facilities. During the storm, Kasim not only protected his property but also rescued and supported his neighbors. And, after the storm, Kasim invested in his business and demonstrated his commitment to the Lower Ninth Ward. As Chamlee-Wright and Storr (2008: 13) conclude, "By reopening his business, he also sent an important signal to other entrepreneurs deciding whether or not to invest in the community and to displaced residents deciding whether or not to return." Furthermore, Kasim notes how his business brought him closer to the community after the storm. "The community is also very supportive," he explains, "they come here, bring their business. It is like a kind of family." Kasim served the community not only by providing needed goods and services but also by establishing a centralized location that became the focal point for community recovery. Using the language from chapter 3, Kasim signaled that he was committed to the community and to its return, and he increased the perceived probability that others would also return. By reopening his convenience store, Kasim was able to signal not only a commitment to the community but also that community recovery was under way.[19]

Inspiring Others to Return by Reopening a Business in Central City

Across town from Kasim was another business owner committed to community recovery, Mike Dean†, who owns a grocery store in Central City. Dean has been in charge of the store for over 30 years, and it has been in his family for over 70 years. He reopened just six months after the storm in February 2006.

Central City is a neighborhood in the lower part of Uptown and just above the business district in the city of New Orleans. Prior to Hurricane Katrina, the 1.4 square mile neighborhood was home to just over 19,000 people. Predominately African American (87.1 percent), most residents were renters, with only 16.3 percent owning their own home. The average household income was $32,021. In "Central City," as a local pastor described, "you had a lot of residents in the area with low income to no income, limited housing. We had some commercial [establishments] in that area."[20] Many people that work in the neighborhood live elsewhere in the city. However, there are several nonprofit organizations and businesses in the area. One nonprofit employee categorized the area as "definitely a hub of activity...There was just a lot of organization, a lot of culture, a lot of history and all of that was really important to people in this neighborhood and continues to be important in this neighborhood."[21]

Although some buildings in Central City received substantial flood damage as a result of Katrina, others received mostly wind damage. For instance, one resident said her roof was torn completely off and another said her house looked as if it had been hit by a tornado.[22] However, even wind-damaged property had mold from being exposed to the elements after the storm. Stewart Martin†, who lived in Central City his entire life and worked for the Central City Housing Development Corporation, described the diversity of damage in the area. As he described, "Yeah...let's say four blocks over, it was just devastated. Eight feet of water. And, four blocks down, no water. It's unimaginable. And, we had one house that blew completely to the ground."

Dean lives in Metairie, three miles from Central City. However, his family has owned the grocery store in Central City for over 70 years, so he has strong ties to the area. Similar to Kasim's view of the Lower Ninth Ward, Dean has a positive view of the neighborhood. While he admits, "The crime is still here. The crime is still here. It's still mixed." He nonetheless explains, "You got your criminals and you got your decent people...There's a lot of good people here. They've made movies around here. They've, it's sad that some of the bad people make it horrible for the good people. There's 90 percent good people, ten percent bad and they ruin it for all of them, to be honest with you."

Dean and his family evacuated before Hurricane Katrina hit New Orleans, first going to Mississippi and then Florida to escape the storm's path. They then returned to Louisiana, staying in Baton Rouge until he returned alone two weeks later. His family stayed in Baton Rouge while he assessed the damage and began to rebuild. Although his house in Metairie had no damage, he noted that in his absence "the business was looted and flooded." So he began the process of rebuilding the store as well as the other properties he and his father owned throughout the area.

While he did not have to wait for insurance payments or assistance to start rebuilding, he did have to find new suppliers, since most of the grocery store's inventory had come from the Ninth Ward, which was completely devastated after the storm. Dean recalled, "Suppliers are very hard to come by. Mostly everything I was getting from the store was coming through the Ninth Ward area. That's where they had the worst hit, which you saw on CNN and everything. That was where a lot of my suppliers were. They're gone now. They're gone now. So I found new suppliers, which are very hard to come by."

Once he reopened the grocery store in February of 2006, it was able to provide critical resources to the community. Furthermore, as Dean recognized, his grocery store signaled rebound to the community. As he explains, "we're like a little heartbeat in the city. We had so many customers that [ask], 'Ya'll open?', and by us saying, 'Yeah, we're open, come on back,' I think that has been a worm on a hook type thing to get people back. It has been instrumental in bringing people back here." Not only did Dean recognize that returning residents would need groceries, premade meals, and supplies, but that they would also need access to ATMs, fax machines, and copiers. These services allowed customers to pay their bills and submit paperwork to insurance companies and FEMA. Dean also allowed his customers to use his fax number as a contact and received and held their messages for them.

The grocery store also served as a meeting place where day laborers could get breakfast, ice, and other supplies and get picked up to work for the day. As Dean noted, "This store is a big social place." Indeed, the array of goods and services Dean provided created a social space for the community, where they would purchase supplies, submit paperwork, and exchange information.

Dean's store was one of the first to open up in the area and was, thus, a crucial focal point for recovery in Central City.[23] He was able to rebuild and reopen quickly, providing a useful signal to other residents and business owners looking to return. Dean also saw an opportunity to provide needed resources and services to returning residents and workers, including access to a fax machine to submit paperwork and receive correspondence.

Dean was able to quickly reopen his grocery store and start providing resources that were desperately needed during the recovery process. In doing so, he created a social space for others to gather, share information, and submit paperwork for insurance money and disaster assistance. By returning to business as usual, Dean signaled to others that return was possible and that community rebound was likely (see Chamlee-Wright 2010). In the language from chapter 3, Dean increased the probability that community rebound would occur by helping disaster victims in his community return to normalcy as quickly as possible.

Coordinating Relief Services in Far Rockaway

The close-knit Orthodox Jewish community in Far Rockaway used their existing organizations and networks to their advantage after Hurricane Sandy. While Achiezer (discussed in chapter 5) was the nexus for coordination for the broader community, other local leaders stepped in to organize response and recovery throughout the community. Three synagogues became resource centers that provided hot meals, clothing, generators, and other needed goods and services. These resource centers allowed residents affected by Sandy to talk to their neighbors, get supplies, and coordinate repairs. One such center was located at the White Shul in Far Rockaway. Chaim Leibtag, the president at the White Shul, faced the challenge of coordinating relief services and securing resources for the synagogue's congregation.

Recall that Far Rockaway is a fairly diverse area on the Rockaway Peninsula in New York.[24] Also recall that there was up to six feet of flooding in some areas as a result of Hurricane Sandy. While some houses and buildings (particularly those further away from the coast and on higher elevation) received little flood damage, others were severely damaged. And, everyone was without utilities and with limited access to food, supplies, and resources for the two weeks following the storm.

By the first weekend after the storm, members of the congregation began gathering at the White Shul to address the pressing issues of a prolonged lack of electricity, supplies, and schooling. According to Leibtag, the synagogue was the logical location to gather since it is the center of the Orthodox Jewish community. Most of the 500 members live within a mile of the White Shul so they can walk there in order to abide by the rules of the Sabbath. For that reason, the synagogue is the literal center of the community in addition to being the spiritual and cultural center.

While Leibtag had lived and worked in the Far Rockaway Orthodox Jewish community for decades, he was new to his position as president of the White Shul when Hurricane Sandy hit. Yet, this did not stop him from working with others in the community to turn the synagogue into a relief center. Within a short time they had a generator and had set up outdoor lighting typically used for holidays. Leibtag also had an Internet connection and charging station hooked up so that members of the congregation could charge their electronic devices and use the Internet to check on family and request supplies and services. Then food and clothing started coming in, including fresh groceries that were donated by a local grocer. Volunteers and residents began cooking hot meals, serving roughly 300 meals three times a day, and setting up space to distribute supplies.

A truck of gasoline was arranged to stop at the synagogue, and Leibtag realized they needed to be organized about the distribution. First, when someone from Maryland offered to pay for a bus to take people who wanted get away down south, Leibtag asked them to return the empty bus with gasoline containers. He then emailed out instructions to the congregation, setting up times to pick up filled containers of gas, giving first priority to emergency personnel and then fulfilling the needs of residents. The distribution went smoothly and provided the needed fuel to run generators and equipment necessary for conducting repairs.

As the adults dealt with relief and rebuilding efforts, the children of the congregation were getting restless. Leibtag worked with some parents and teenage volunteers to set up activities, lessons, and entertainment. They held classes and organized crafts and shows by musicians and even a magician. These activities went well into the evening each night and were a welcome break for the adults who were dealing with clearing debris, draining floodwater, and repairing their homes.

Leibtag then worked with a local pediatrician to set up a clinic in the White Shul. Dr. Hylton Lightman's office had suffered over five feet of floodwater and sewage. Fortunately, Dr. Lightman and his wife had prepared for the storm by backing up their electronic records, which allowed them to easily set up shop in a temporary location. As Dr. Lightman (2013) recalled, "Within seventy-two hours of Mr. Leibtag's offer, we were fully operational." Dr. Lightman ended up staying at the White Shul for six months while his office was gutted and rebuilt.

When FEMA arrived two weeks after the storm, they set up an information center in the White Shul since it was the place where residents came for food, resources, and information. According to Leibtag, around two weeks into recovery the National Guard came by and offered a truck load of food. While the food was Kosher, Leibtag realized they did not need additional food and gave it to a local church.

Leibtag also organized the White Shul into a relief center that provided needed goods and services for the congregation. As a natural gathering place in normal times, the synagogue quickly became the gathering place after Hurricane Sandy. In a testimony at a New York City council hearing on the storm, Leibtag (2013) pointed to the central importance of religious institutions in times of crisis, "Synagogues, mosques, and churches are the centers of their communities. They become that way because that is the nature of what they are. Even the unaffiliated and non-believers come to a house of worship in difficult times because they know deep inside that this is the place where they will receive help."

By being flexible and utilizing his connections in the community, Leibtag ensured that the congregation had the support they needed to recover. Leibtag (2013), again in his testimony, proudly stated that he "had the privilege to lead my synagogue during the critical weeks following the storm as we provided food (three meals a day), shelter, electrical power strips and Wi-Fi, gasoline and critical information to those impacted." Not surprisingly, his community has recognized him for his efforts. For instance, Met Council, a Jewish human service organization in New York, awarded Liebtag along with others in Far Rockaway for going above and beyond after the storm.[25]

Leibtag not only ensured that the congregation had the goods and services they needed but also worked to make sure they received the information they needed, as well as some rest and companionship in the weeks following the storm. In the language we introduced in chapter 3, Leibtag reduced the costs of recovery by helping to provide immediate relief efforts and, in the process, signaled that recovery was under way.

Leibtag, like Rabbi Kruger and Mrs. Russell (discussed in chapter 6), utilized his experiences and connections to provide needed goods and services after Hurricane Sandy. He was able to connect aid suppliers with demanders by opening up the space of the White Shul and coordinating food, gasoline, and clothing distribution, as well as providing space for a doctor to see patients. His efforts echo the importance of local knowledge and community leaders in providing assistance after a disaster (Sobel and Leeson 2007; Chamlee-Wright and Storr 2008, 2009c, 2010a, 2014; Horwitz 2009a).

In Summary, Entrepreneurs Act as Focal Points for Recovery after Disasters

The above examples highlight the importance of entrepreneurs signaling that community rebound is likely and, in fact, under way. Father Vien not only encouraged his parishioners to return to New Orleans East after Hurricane Katrina but also ensured they had the resources needed to return. Doris Voitier recognized that a functioning school system was necessary to entice families to return and navigated and circumvented the bureaucratic process to fulfill her promise of providing schooling for residents of St. Bernard Parish. Casey Kasim saw a profit opportunity in opening a gas station, convenience store, and laundry facilities in the Lower Ninth Ward. In the process, however, his store became an important social space in the community, both before and after the storm, and his store's reopening signaled a return to normalcy. Mike Dean's store in Central City was also an important social space after Hurricane Katrina, and in reopening it he signaled to residents that recovery was likely, and that there would be a place to obtain food, supplies, and information needed to rebuild. Additionally, the White Shul in Far Rockaway, and its president Chaim Leibtag, provided resources and support for its members, most of whom lived within walking distance to the shul. While each entrepreneur discussed in this chapter fulfills his or her role differently, the common characteristic is their ability to signal return, rally supporters, and organize recovery.

Chapter 8

Fostering Resilient Communities

Entrepreneurs, as we discussed in chapter 2, recognize and pursue opportunities to change the world. For instance, they bring new products or services to the market, such as offering an automobile that has a bundle of features other cars do not have or starting a landscaping business in an area currently underserved, in the hope that potential customers will find these new products or services valuable. Similarly, entrepreneurs start social enterprises in an attempt to solve social problems, such as opening an after-school program for troubled teenagers or organizing a petition to change a city ordinance to prevent the dumping of trash in a particular area. They undertake these social enterprises in the hope that the lives of community members will improve and potential donors and volunteers, who are also concerned about these problems, will believe these social enterprises are helping to solve them. Given that the future is unknowable, an entrepreneur's hopes could prove to be overly optimistic, in which case his enterprise will not succeed. If an entrepreneur's hopes prove to be well-founded, however, she will provide goods and services that people actually desire, she will earn profits and/or receive donations, she will attract employees and/or volunteers, her organization will thrive, and she will advance broader social change. Purchasing landscaping services might allow working parents to spend their Sunday afternoons engaging with their children rather than caring for their lawn. Participating in an after-school program might alter the life of a disadvantaged student, increasing the likelihood she will obtain a college education and pursue a career she previously believed was beyond her reach. Entrepreneurs are, thus, social change agents who, despite the radical uncertainty we all necessarily confront in the world, notice, cultivate, and exploit opportunities to bring about economic, social, political, institutional, ideological, and cultural transformations.

As we discussed in chapter 3, the uncertainty entrepreneurs confront in mundane times is even more pronounced after a disaster. After a disaster, for instance, the critical infrastructure and public services that entrepreneurs rely on are likely to be disrupted, and the time frame of their restoration is likely to be in doubt. Similarly, many entrepreneurs' employees, volunteers, followers, suppliers, customers, clients, and/or stakeholders are likely to have been affected by the disaster, they may even be displaced and the time frame for their return might be in doubt. Additionally, an entrepreneur might have also suffered damage to his home and the buildings that host his enterprise, and his own calculus concerning whether or not to reopen or relocate his enterprise might be complicated. And, since community rebound after a disaster is a collective action problem, community recovery is likely to be slow and uncertain.

Entrepreneurs, however, as we argued in chapter 3, are well placed to help communities overcome the collective action problem that characterizes post-disaster situations. Just as they drive social change in mundane times, entrepreneurs can promote post-disaster community rebound by (1) providing needed goods and services, (2) helping to restore and replace disrupted social networks, and (3) signaling that community rebound is likely and, in fact, under way. The entrepreneurs whose efforts we discussed in chapters 5, 6, and 7 demonstrate that entrepreneurship after disasters can help communities overcome the post-disaster collective action problem. For instance, Alice Craft-Kerney, discussed in chapter 5, was alert to the pressing health needs in the Lower Ninth Ward and opened a clinic for residents who were trying to recover from Hurricane Katrina. Rabbi Kruger, discussed in chapter 6, helped his neighborhood recover quickly after Hurricane Sandy by collecting a detailed list of needed items for aid suppliers to fulfill and, thereby, connecting aid suppliers with aid demanders. And, Father Vien, discussed in chapter 7, personally encouraged his parishioners who were displaced by Katrina to return to New Orleans East and, once they returned, organized their recovery efforts and advocated for resources on their behalf.

The notion that entrepreneurs can drive post-disaster community recovery may be surprising given that governments are often assumed to be the necessary drivers in disaster mitigation, management, and recovery efforts (see, for instance, Birch and Wachter 2006; Schneider 2008; Thaler and Sunstein 2008; Cigler 2009; Springer 2009, 2011). Private actors would appear to lack the resources to deal with the scale of destruction caused by disasters. Additionally, private actors can often appear to lack the coordinative capacity to deal with the scope of the coordination challenges presented by post-disaster preparedness, relief, and recovery. Our analysis, however, does not seek to challenge the importance of government involvement in disaster mitigation, management, and recovery efforts. In fact, by highlighting the role of the entrepreneur in post-disaster recovery but recognizing that entrepreneurship occurs in private and public settings, we hope to move the conversation away from discussions of whether private or public responses to disasters are most likely to be effective at promoting community rebound. Instead, by pointing to the critical role of entrepreneurs, we hope to move the conversation toward discussions of whether entrepreneurs are given adequate space to act in post-disaster environments or are handcuffed in their efforts to promote recovery.[1] As such, we contrast polycentric approaches to disaster management and recovery with monocentric approaches. We argue that approaches to community resilience and disaster recovery that adopt principles of polycentricity, and so ensure there is space for entrepreneurial action, are more likely to encourage community rebound than monocentric approaches.

Vincent Ostrom Draws a Meaningful Distinction between Monocentrism and Polycentrism

Vincent Ostrom, mentioned in chapter 3, has distinguished between systems organized according to monocentric principles and systems that are characterized by patterns of polycentric ordering. There is a unity of power within a monocentric system, with decision rights centralized in the hands of a single decision-making authority. In such a political system, the power to implement and enforce rules is "vested in some single office or decision structure that has an ultimate monopoly over the legitimate exercise of coercive capabilities" (V. Ostrom [1972] 1999: 55). As introduced in V. Ostrom et al. (1961), a polycentric political system, on the other hand, is one comprised of (1) multiple autonomous centers of authority that are, at least formally, independent of each other but, nonetheless, (2) take each other into account as they make decisions, as well as (3) cooperate with and compete against each other. V. Ostrom ([1991] 2014: 48) has explained that it is possible to identify patterns of polycentric orderings in a variety of settings, including "(1) competitive market economies, (2) competitive public economies [like the constellation of municipal governments in the US], (3) scientific inquiry, (4) law and adjudicatory arrangements [in Western countries], (5) systems of governance with a separation of powers and checks and balances [like the US federal government], and (6) patterns of international order." The "whole system of human affairs" may also be described as exhibiting patterns of polycentric ordering (ibid.: 55). As noted earlier, monocentric systems are the space of the bureaucrat, who reinforces the status quo, and polycentric systems are the space of the entrepreneur, who promotes social change.

It is important to note that, in any actual system of social order, we are likely to find elements of both monocentrism and polycentrism. As V. Ostrom ([1972] 1999: 55) writes, "the existence of a predominantly polycentric political system need not preclude elements of monocentricity from existing in such a system." Likewise, "a predominantly monocentric political system need not preclude the possibility that elements of polycentricity may exist in the organization of such a system" (ibid.: 55). And, "aspects of polycentricity are likely to arise in all systems of social order because human beings are capable of thinking for themselves and acting in ways that take account of their own interests" (V. Ostrom [1991] 2014: 46). It is not uncommon to see bureaucratic organizations where decision-making authority is centralized but exists within broader polycentric orders. Similarly, it is not uncommon to see some entrepreneurial elements in otherwise monocentric systems.

Polycentric systems, however, promote self-governance in a way that monocentric systems do not. As V. Ostrom ([1991] 2014: 46–47) explains, "polycentricity serves as a structural basis for the emergence of actual self-governing arrangements," and, "[t]he autonomous character of polycentric systems implies self-organizing capabilities." Additionally, as Aligica and Sabetti (2014: 9) suggest, "polycentrism may generate a very special governance structure that fosters self-governance." Also, as McGinnis (1999a: 2) writes, "only polycentric governance can nurture and sustain the self-governing capabilities of local communities." Indeed, local communities appear to be willing and able to develop self-governing arrangements to effectively deal with collective challenges. Elinor Ostrom's (1990) work on governing the commons, for instance, suggests that communities are quite capable of developing mechanisms and procedures for limiting access to common-pool resources and avoiding tragedy of the commons problems (see also McGinnis 1999a, 2000; Basurto and E. Ostrom 2009; Jack and Recalde 2013; van Laerhoven and Barnes 2014). According to McGinnis and Walker (2010: 294), "the basic idea [of polycentricity] is that any group of individuals facing collective problems should be able to address that problem in whatever way they best see fit." Polycentricity recognizes that there are advantages to allowing complex problems to be solved at the community, group, neighborhood, and individual levels.[2]

Arguably, polycentric systems are likely to be better than monocentric systems at solving community challenges where flexibility and adaptability are important, such as during times of crisis. As such, polycentric systems, because they give entrepreneurs room to act, are likely to outperform monocentric systems when flexibility and adaptability are important. This is certainly true in the competitive market, the quintessential polycentric order, where entrepreneurs effectively confront an uncertain world but nonetheless provide consumers with goods and services they desire and, as judged by their willingness to voluntarily engage in exchanges, improve the lives of their customers (Smith [1776] 1981; V. Ostrom [1972] 1999). The political system of federalism, like the one that exists in the United States, is another example of a polycentric order that deals adequately with complex challenges and is able to weather crises (V. Ostrom [1972] 1999; Krane 2002).

The post-disaster situation is a circumstance where a great deal is in flux. Consequently, the most effective approaches to disaster management and recovery will likely be ordered according to the principles of polycentrism. In the post-disaster context, for instance, local governing units with the authority to make decisions are more likely than centralized governing authorities to overcome the challenges associated with disaster management and recovery because of their access to local knowledge as well as their greater flexibility and adaptability. Coyne and Lemke (2011, 2012) have highlighted two distinct benefits of a polycentric system in disaster relief. First, polycentric orders allow for flexibility to provide disaster relief at various scales. "Perhaps the most compelling case for dispersing decision making power across multiple autonomous groups," as Coyne and Lemke (2012: 223) write, "is the flexibility of polycentrism to adapt to different scales of public goods provision." Polycentric orders allow community problems to be solved at the level of society that is best situated to solve the problem. Monocentric systems, conversely, risk closing off local solutions to complex challenges like disaster recovery. Second, polycentric orders tend to make more effective use of dispersed knowledge after a disaster. As Coyne and Lemke (2012: 223) write, "polycentric orders have some distinct advantages…[including] the ability of decentralized organizational structure to capture the specific knowledge of local actors." And, "the fewer the layers of bureaucracy between consumer and producer, the more grounded the producer's decisions will be in terms of understanding the consumers' wants and the value placed on the provision of the good" (ibid.: 223).

While polycentric orders increase the social capital and coordination capacity of community organizations and associations by allowing them to utilize local knowledge as they respond to challenges, monocentric orders tend to crowd out local efforts to respond to challenges (E. Ostrom 2000, 2014). This has implications for understanding the challenges associated with disaster mitigation, management, and recovery, as well as the policies that are likely to foster community resilience.

Monocentric Systems Cannot Adequately Solve the Post-Disaster Collective Action Problem

As discussed in chapter 3, monocentric systems are not well positioned to overcome the collective action problem after a disaster. In a monocentric system, all decisions are funneled toward the top, and all directions are funneled downward in the organizational hierarchy. This process is more complicated the more layers that exist within the hierarchy. The more layers of hierarchy or the greater the distance between the decision maker and the customer, client, or stakeholder, the more complex and error-prone the messages will become (Tullock [1965] 2005). Also, the greater the distance, the longer it will take for the decision maker to learn about errors in implementation or changes in customer preferences, further complicating any ability to fix errors and adapt to changing circumstances. As V. Ostrom and E. Ostrom ([1971] 2014: 32) explain, "[L]arge-scale bureaucracies will, thus, become error prone and cumbersome in adapting to rapidly changing conditions. Efforts to correct the malfunctioning of bureaucracies by tightening control will simply magnify the problem."

Monocentric systems operate as centralized, top-down decision-making institutions, using formal protocols and preestablished procedures to assess alternatives and to respond to problems (Weber 1978; Mises [1944] 2007; Tullock [1965] 2005). While this rigid structure may allow for a large number of people to cooperate to solve particular large-scale problems, it also restricts the ability of the order and the individuals within it to adapt to unforeseen or changing circumstances. Thus, the bureaucratic structure can result in the implementation of rigid and inflexible policies by even the most well-intentioned policymakers.[3] For instance, education policy in the United States has become more and more centralized in attempts to ensure that all children have access to affordable schooling and to maintain a standard of education across the nation. In doing so, procedures for selecting and implementing curriculum, teaching methods, and testing have necessarily been centralized through policies such as "No Child Left Behind." While these efforts allow for measuring certain comparable outputs of education, they have severely hindered the ability for individual schools and teachers to tailor education to the particular needs of their students (see Hanushek 2003; Darling-Hammond 2007; Guisbond and Neill 2004).

Moreover, monocentric systems cannot adequately overcome the problem associated with collecting, deciphering, and acting upon dispersed and inarticulate knowledge in mundane times. They are likely to be even more epistemically challenged after a disaster. Recall the collective action problem that exists after a disaster, which was extensively discussed in chapter 3. Infrastructure, homes, and businesses are damaged or destroyed, utilities and other public services are suspended, and residents are dispersed by disasters. Everyone affected by the disaster has limited access to information. Additionally, everyone is uncertain about what to do next, since the course of action that makes the most sense for each disaster victim is partly dependent on what others decide to do. In this situation, as we demonstrated, it might be rational for everyone (including bureaucrats responsible for disaster response and recovery) to adopt a wait-and-see strategy. Residents, for instance, may want to know that utilities have been restored and schools reopened before committing to return and rebuild. Likewise, bureaucrats may want to see that people are willing to return before investing in repairing infrastructure and reestablishing services.

Even if bureaucracies had easy access to unlimited resources, they would still need to overcome the knowledge problem to effectively lead relief and recovery efforts. Sobel and Leeson (2007), for instance, have argued that effective disaster relief must be able to determine (1) that a crisis has occurred, (2) what needs must be met and how resources should be allocated to best serve those needs, and (3) how to evaluate and adjust the relief effort as circumstances change. By examining the relief effort after Hurricane Katrina, they determined that federal government response failed on all three measures (ibid.). Similarly, effective efforts to promote recovery must determine (1) which residents are most likely to return and which neighborhoods are most likely to rebound, (2) how best to allocate resources, and (3) when it has made mistakes and how best to correct them (Storr and Haeffele-Balch 2012). Again, centralized, government-led efforts have proven to be inadequate in these aspects (see Chamlee-Wright and Storr 2009c, 2010b). Approaches to disaster recovery that exhibit monocentric principles have invariably hampered decentralized recovery efforts. One example of this was the series of redevelopment plans introduced after Hurricane Katrina, each one with different, and sometimes competing, information about which areas should be allowed to rebuild and which should not (Olshansky et al. 2008). Recall that LaToya Cantrell, discussed in chapter 6, and members of the BIA invested considerable energy in proving that they were a viable community.

Another example was the restrictive stance toward occupational licensing adopted in New Orleans after Katrina (Skarbek 2008); and another issue was the difficulties that residents and businesses experienced in obtaining building and business permits (Chamlee-Wright 2010). Recall that Alice Craft-Kerney was delayed in opening her clinic until proper permits were obtained, and that Father Vien had to lobby officials to permit him to establish a FEMA trailer park on church grounds.

While policymakers can encourage the recovery of certain communities by offering various incentives or providing needed resources, absent information on the plans and purposes of the myriad affected and displaced residents, they lack much of the knowledge needed to make a nonarbitrary decision regarding which communities to encourage and which to discourage.

Instead, polycentric systems, which allow considerable space for entrepreneurial activity, are more likely to overcome the collective action problem that complicates post-disaster recovery. As discussed in chapters 2 and 3, entrepreneurs are alert to opportunities to improve upon existing circumstances and to offer new products, services, or ideas to promote social change in both mundane times and after disasters. Entrepreneurship is broadly conceived here to include not just commercial entrepreneurs but also social and ideological entrepreneurs. It would also include political entrepreneurs (i.e., local government officials with the autonomy and responsibility to act, the positioning to reap the benefits of a job well done, and the likelihood that they will pay the cost of failure).[4] For us, the category includes business owners like Mary Ann Patrick, Casey Kasim, and Mike Dean, who worked to reopen their businesses, as well as Ben Cicek, who opened a new store, to provide needed goods and services. The category also includes social activists and leaders such as Rabbi Bender, Father Vien, and LaToya Cantrell, who adapted their preexisting organizations, as well as Alice Craft-Kerney, who opened a new clinic to address the new needs of their community. Additionally, the category includes government officials, such as Doris Voitier, who took on the challenge of opening schools in St. Bernard Parish and stuck to her promise by relying on her own people and networks when official channels proved too burdensome.

Unfortunately, policymakers often propose and implement policies that hamper entrepreneurship after disasters. Policymakers should, instead, propose and implement policies that ensure entrepreneurs have the space to act.

Policymakers, However, Sometimes Favor Monocentric Approaches to Disaster Recovery and Pursue Policies or Act in Ways That Hamper Entrepreneurship and Community Recovery

Indeed, policymakers sometimes adopt policies or engage in practices that not only worsen the post-disaster collective action problem (by increasing the costs associated with rebuilding and decreasing the benefits of return) but also interfere with the ability of entrepreneurs to assist in recovery.

Policymakers can constrain entrepreneurial activity and slow community rebound by adding to the uncertainty that is endemic to the post-disaster environment and dampening an entrepreneur's ability to identify opportunities and pursue their plans. There is, in fact, a considerable literature on how policymakers can raise the cognitive burden that community members must contend with after a disruptive event.[5] Higgs (1997), for instance, examined the slow recovery in the United States after the Great Depression and found that persistent regime uncertainty caused the public to lose faith in the government's commitment to protect private property rights and to uphold the rule of law, and this made it difficult for individuals to form reasonable expectations about the future. When policies are in flux, individuals find it difficult to form reasonable expectations about the future and may wait until the circumstances are more certain before forming plans and investing in future endeavors.

Building on Higgs' analysis, Chamlee-Wright (2007) argues that the post-Katrina environment was rife with regime uncertainty, which complicated the ability of disaster victims to form reasonable expectations about what the government was likely to do. The signal noise that occurs when policymakers delay and reverse decisions and when they send mixed and confusing messages about which policies they will adopt also complicates the ability of entrepreneurs to notice and pursue opportunities to meet the needs of disaster victims. As Chamlee-Wright (ibid.: 240–241) observed:

> [I]n the course of our team's investigations of community rebound, we encountered time and again people frustrated by the confusion and uncertainty in decision contexts. Their frustration arose not simply because there was much to learn, but because they had to waste time learning about things that ultimately seemed superfluous to the recovery effort, such as navigating the bureaucratic maze of relief agencies and regulatory policy. Worse yet was the sense of frustration they felt when they realized that no amount of effort on their part would reduce the uncertainty they faced because the institutional rules of the game seemed always to be in a state of flux.

While some disaster victims will still push forward with their plans to return despite the noise, others may relocate elsewhere even if they would have preferred to return. Similarly, while some entrepreneurs will still pursue their plans, other potential entrepreneurs will abandon their attempts to meet the needs of disaster victims.

The recovery planning process and the monocentric system of disaster relief were major sources of signal noise after Hurricane Katrina (Chamlee-Wright 2010). After Hurricane Katrina, redevelopment plans only added to the uncertainty, both because they often required residents to prove that their neighborhoods were worth rebuilding and because the plans were constantly being changed (ibid.). Louisiana's Road Home Program, which was designed to assist disaster victims in rebuilding their damaged homes, also contributed to the signal noise disaster victims experienced after Katrina. Although the program promised assistance to disaster victims who wanted to rebuild, the program was poorly administered. Two years after Katrina, only a tiny fraction of the applications, 18 percent, had been closed (Pike 2007). Additionally, there was considerable uncertainty surrounding the size of the award that disaster victims could expect to receive, as well as confusion about the actual process of applying for assistance (Chamlee-Wright 2007, 2010).

Signal noise also negatively affected several of the entrepreneurs that we discussed in chapters 5, 6, and 7. LaToya Cantrell and the other residents of Broadmoor in central New Orleans, for instance, were shocked to find out that their neighborhood was initially designated to become a green space, according to the BNOB Commission's plan. They then found out that they would need to prove their neighborhood's viability (i.e., that at least 50 percent of residents were committed to return) before the plan would be revised. Instead of focusing primarily on rebuilding their neighborhood, Broadmoor residents had to organize a petition to stop the "green dot." Similarly, Father Vien's community in New Orleans East faced uncertainty from the redevelopment plans as well as issues with getting utilities restored. In response, Father Vien compiled a roster of people who had attended mass at the church and prepared a photo album of residents to prove a need for restoring services. Because of signal noise, Father Vien had to invest his time in advocating for the community rather than on actual rebuilding. Ed Williams, who worked as a contractor before and after Hurricane Katrina, had to deal with the uncertainty and delays surrounding disaster assistance payments from FEMA and the Road Home Program. Specifically, when his clients did not receive assistance checks on time, he would have to wait to receive his payments as well. This meant that he often had to operate with negative cash flows until his clients received assistance. He also could never really be sure that he would be paid for his efforts. Additionally, other entrepreneurs faced issues with permits, including Mary Ann Patrick, Alice Craft-Kerney, and Casey Kasim.

In addition to adopting policies or engaging in practices that add to the uncertainty that is endemic to the post-disaster environment, policymakers sometimes adopt policies that dampen, distort, or destroy the signals entrepreneurs rely on to guide their actions. Specifically, policymakers sometimes directly interfere with post-disaster entrepreneurship by subsidizing certain activities while raising the costs of others.

Post-disaster subsidies funnel activities into particular directions and can lead to a misallocation of resources. For instance, extending unemployment benefits for months after a disaster may help disaster victims get back on their feet, but the longer those benefits are in place, the higher the incentive to refrain from making firm plans for recovery and returning to normalcy. Thus, extending unemployment benefits can make it more difficult for entrepreneurs to find employees. Similarly, the influx of short-term employment opportunities with disaster management organizations can entice some residents to delay their search for long-term employment. If low-skilled workers can earn premium wages by working for a disaster management agency, entrepreneurs may not be able to attract needed workers (Chamlee-Wright 2010).

Post-disaster price controls can also limit entrepreneurial discovery and social learning. For instance, many local and state governments implement price gouging laws to prevent business owners from dramatically increasing the prices of certain scarce items, including gasoline and ice. While such laws may try to stop people from profiting off of the losses of others, they also distort the price signals that are essential to allocating resources to their most valued uses.[6] When prices are prevented from rising, fewer people will find it worthwhile to send supplies to the affected area (while some entrepreneurs may still provide the needed goods or services, others will not). Thus, shortages will occur, resulting in the need for disaster victims to stand in line for hours to purchase gasoline, ice, or other needed supplies (for instance, see Brewer 2006–2007).

Signal dampening, distortion, and destruction affected several of the entrepreneurs that we discussed in chapters 5, 6, and 7. For instance, Mary Ann Patrick had to fight with FEMA and Parish authorities to prove that her business property was not condemned. In order to do so, she had to work with her insurance company and hire a structural engineer. Patrick's time and resources were diverted to dealing with property assessment rather than to actually reopening her business. Other entrepreneurs, such as Father Vien and Doris Voitier, also had to spend time learning about the disaster assistance process. While these activities resulted in funds flowing to their communities, they often came with stipulations that dampened their effectiveness (Chamlee-Wright 2010). For instance, both Father Vien and Voitier had to get additional permission to set up trailers after they were acquired.

Returning to the model of the post-disaster collective action problem introduced in chapter 3 (see figures 3.1 and 3.2), there are several ways policymakers can change the nature of the game or affect the payoffs that disaster victims receive. In chapter 3, we discussed how policymakers could positively impact recovery. There are also several ways that policymakers can hinder, distort, or dampen community rebound. For instance:

- Policymakers can delay return by closing off neighborhoods for extended periods of time (i.e., decreasing P_r). Policymakers can also decrease the perceived likelihood that others will return (i.e. decrease P_r) by adding signal noise to the recovery process, such as engaging in an extended redevelopment planning process and delaying decisions regarding flood maps and insurance requirements.
- Policymakers can decrease the benefits associated with returning by delaying the restoration of public goods (i.e., introducing a negative θ). Also, they can implement policies that dampen entrepreneurial action, such as price controls, occupational licensing requirements, and extended unemployment benefits (i.e., eliminating the possibility of adding θ by preventing a new good or service from entering the market).
- Policymakers can increase the cost associated with returning by offering to subsidize rebuilding but then implementing a complicated and uncertain application and delivery process (i.e., introducing a negative γ). They can

also divert resources that could have been spent on recovery by requiring disaster victims to prove viability and to fight policies that would condemn their homes and businesses (i.e., introducing a negative γ).

This list is clearly not exhaustive. But, taken together, they suggest that it is possible for policymakers to alter the game, making recovery more costly, difficult, and uncertain. Such policies also hinder entrepreneurial action by distorting the signals that encourage entrepreneurs to act. Arguably, the more monocentric the approach to disaster management and recovery, the more likely it is that policies that hamper entrepreneurship will be adopted, the more dismal their consequences are likely to be, and the less likely it is that these policies will be modified or abandoned.

Policymakers Should, Instead, Embrace Polycentrism and Ensure That Entrepreneurs Have the Space to Act

It is through the dedication and success of entrepreneurs that communities flourish. As the examples in chapters 5, 6, and 7 have highlighted, this is true in both mundane times and after disasters. Arguably, creating resilient communities also means embracing polycentrism and creating communities where entrepreneurs have the space to act.

There is considerable literature available on resiliency. Resilient communities are communities that not only survive disasters but also thrive after disasters.[7] Holling (1973: 14), for instance, has described resilience as "a measure of the persistence of systems and of their ability to absorb change and disturbance and still maintain the same relationships between populations or state variables." As Kulig (2000: 375) notes, community resilience is "the ability of a community to not only deal with adversity but in doing so reach a higher level of functioning." Adger (2000: 361), likewise, describes resilience as "the ability of communities to withstand external shocks to their social infrastructure." Similarly, Walker et al. (2002: 14) defined resilience as "the potential of a system to remain in a particular configuration and to maintain its feedbacks and functions, and involves the ability of the system to reorganize following disturbance-driven change." And, Paton (2006: 8) likewise explains, "resilience is a measure of how well people and societies can adapt to a changed reality and capitalize on new possibilities."[8] To say that a community is resilient is to say that it has adaptive capacity as well as the ability to learn from disasters and to make the most of the possibilities for positive change and development that exist after disasters. Community resilience is, thus, related to concepts like "community competence" (Pooley et al. 2006; Brown and Kulig 1996–1997) and "successful neighborhoods" (Jacobs 1961).

Being resilient does not merely mean being able to return to a previous condition after experiencing some shock. Although the Latin root of the word resilience, *resiliere,* means "to jump back," as Paton (2006: 7) explains, "this usage . . . captures neither the reality of disaster experience nor its full implications." First, "jumping back" to a pre-disaster state is not likely to be possible. Second, a disaster might present a community with opportunities to move to a state that would be considered to be preferable to its pre-disaster state by community members. As Paton (ibid.) describes, "even if people wanted to return to a previous state, changes to the physical, social and psychological reality of societal life emanating from the disaster can make this untenable." And, a "disaster can be conceptualized as a catalyst for change; if mother nature does the demolition work, society can make choices about how to rebuild itself" (ibid.: 8). Similarly, as Pooley et al. (2006: 163) describes, a resilient community "does not just bounce back from adverse situations; the community actively chooses change, despite any limitations the community may possess." Resilient communities, then, are communities that rebound or recover from disasters. However, by community rebound and recovery we do not mean that it returns to its pre-disaster state, but that, despite the disaster, the community remains or will again become well-functioning (i.e., a place where community members can thrive).

Communities, it has been argued, do not become resilient by accident.[9] Communities, it has been suggested, must plan for resiliency.[10] They must develop and adopt disaster mitigation and management plans. They must set aside the resources they will need for recovery if a disaster were to occur. They must ensure that community members and community-based institutions are prepared to weather the disaster and have the competencies that the community will have to utilize in order to recover.[11] According to Paton (2006: 9), "neither a capacity to adapt nor a capacity for post-disaster growth and development will happen by chance." Of course, developing a hazard warning system, or advancing building and land use codes that ensure structures are less vulnerable in the face of hazards, or creating emergency relief funds and agencies responsible for disaster management, requires affirmative action from some individual or group of individuals. Moreover, it is understandable to assume that governments can and should be undertaking efforts to make their communities more resilient. But it would be a mistake to assume that it is possible to plan and prepare for all the contingencies associated with disasters. And, it would be an error to assume that entrepreneurs, who are alert to disaster victims' needs and attempt to satisfy them, will not have a key role to play in driving recovery, regardless of how detailed the contingency plans are and how prepared communities appear to be. Although the resilience literature has not tended to emphasize the role or importance of entrepreneurship in post-disaster community rebound, it is difficult to imagine a community that can recover from a disaster that does not give its entrepreneurs the space to act.[12]

Of course, not all entrepreneurs should have the space to act. For instance, the entrepreneurial looter is someone whose efforts should be blocked or discouraged. Instead, it is the entrepreneurs who fulfil the roles necessary for recovery—such as providing goods and services, restoring and replacing social networks, and signaling that recovery is under way—who should be encouraged.

Recognizing that community resilience does not happen by chance and that community rebound might require government's involvement, then, does not give policymakers license to reengineer a community after a disaster. Ambitions to build back the community better and smarter than before are, thus, arguably misguided and are likely to impede the recovery process. They not only delay rebuilding in order to conceive and implement the "right" plan (think of the numerous redevelopment plans that were circulated in the years following Hurricane Katrina), but also worsen the collective action problem by anchoring negative expectations in the minds of those waiting on the sidelines for signs of recovery before they return (Chamlee-Wright and Storr 2009a). Adopting a detailed redevelopment plan does not ensure community rebound after a disaster. Instead, it is the community members themselves and their experiences, relationships, and ambitions; and, most importantly, their entrepreneurial endeavors; that spur recovery.

Once it is recognized that the entrepreneur is a driving force of community recovery after a disaster, it becomes clear that policymakers should ensure that entrepreneurs have the space to act. Specifically, policymakers should adopt polycentric over monocentric approaches to disaster mitigation, management, and recovery; including (1) ensuring that even bureaucrats have the freedom and incentive to act entrepreneurially to meet disaster victims' needs and (2) eliminating, suspending, or simplifying the rules that hamper post-disaster entrepreneurship.[13] These policy recommendations are consistent with the recommendations offered by Chamlee-Wright (2007, 2010), Chamlee-Wright and Storr (2008, 2009c), and Chamlee-Wright et al. (forthcoming).

Ensure that Even Bureaucrats Have the Freedom and Incentive to Act Entrepreneurially to Meet Disaster Victims' Needs

We have already highlighted the differences between bureaucrats and entrepreneurs. While entrepreneurs are agents of change, bureaucrats are preservers of the status quo. While entrepreneurs advance new ways of doing things, bureaucrats are rule followers. If dynamism is associated with the entrepreneur, conservatism is associated with the bureaucrat. Each bureaucrat occupies a particular position within an organizational hierarchy, is expected to possess a particular range of skills and expertise, and is given the authority to perform certain functions within the organization so long as he follows the appropriate rules and procedures. Entrepreneurs, on the other hand, might not be formally connected with any organization or might hold any of the formal positions within a bureaucratic organization (e.g., owner, founder, manager, mid-level supervisor, or low-level staffer). When an entrepreneur is not formally connected with an organization, the laws of the land, the resources that he can attract and command, and any informal constraints he might adopt are the only limits on the range of activities he can attempt. When entrepreneurs are formally a part of a bureaucratic organization, however, their ability to act entrepreneurially likely means that they must act outside of the bureaucratic rules and procedures that proscribe their formal role.

Several of the entrepreneurs that we described in chapters 5, 6, and 7 were formally bureaucrats but acted as entrepreneurs in that they noticed opportunities to serve the needs of disaster victims and attempted to do so in ways that moved beyond their bureaucratic responsibilities. For instance, Doris Voitier is the superintendent of the St. Bernard Parish Unified School District. Her position is within the hierarchy of publicly provided education. However, Voitier stepped outside her traditional role to ensure that schools in St. Bernard Parish would reopen quickly. Whenever she faced issues with obtaining equipment, such as classroom trailers, she got them herself and asked for reimbursement later. Whenever she faced restrictions, such as when she sought to set up FEMA trailers on school grounds for teachers to live in, she petitioned FEMA and the local authorities until permission was granted. By being alert to the needs of the community and personally taking on risks of recovery to fulfil her promise of reopening the schools, Voitier became an entrepreneur. Furthermore, Father Vien, Rabbi Kruger, and Chaim Leibtag all stepped outside their traditional roles as religious leaders and became advocates for community recovery; Sandra Green stepped outside her role as store manager to create a social space for residents, and LaToya Cantrell did more than just fulfil her position as president of the Broadmoor Improvement Association. We describe them as entrepreneurs because of their activities in the post-disaster context, not because of their job titles.

Additionally, several organizations gave their managers the license to be entrepreneurial after Hurricane Katrina. Horwitz (2009a, 2009b), for instance, has highlighted the success of a subset of large companies (such as Wal-Mart and Home Depot) and federal agencies (such as the US Coast Guard), typically known for their hierarchical and bureaucratic nature, in the post-disaster context. The big-box stores had developed business continuity plans, employed staff to track the weather and product shipments, and were able to send employees from other areas to the location of the disaster. Importantly, they also gave autonomy to their middle managers to make decisions on the ground. As Horwitz (2009a: 516) explained, "As Katrina approached, Wal-Mart CEO Lee Scott sent a message directly to his senior staff and told them to pass it down to regional, district, and store managers: 'A lot of you are going to have to make decisions above your level. Make the best decision that you can with the information that's available to you at the time, and, above all, do the right thing' [Rosegrant 2007a: 5]. In Katrina's aftermath, senior management's commitment to this principle was often put to the test." Further, as Horwitz (2009b: 97–98) notes, "The then-vice admiral of the Coast Guard pointed to that autonomy as a reason it was able to move personnel and equipment into place so much more quickly than other agencies during the response to Katrina." This sort of flexibility and decentralization allowed the managers and staff of Wal-Mart and the US Coast Guard to act entrepreneurially and proved that their efforts were essential to the success of their organization's relief efforts after the storm.

Given the challenges associated with post-disaster recovery, policymakers should encourage bureaucrats, wherever possible, to act entrepreneurially after disasters. This does not mean giving every bureaucrat the freedom to ignore all rules that would proscribe their actions and to simply do as they wish. But it does mean encouraging certain bureaucrats (perhaps the most senior authorities in any particular location or middle managers that work in the communities affected by the disaster) to do what they think is best, so long as they behave in ways that are consistent with the overall mission of the organization.[14] There are, of course, limits to this approach. Polycentricism, as it were, is not a panacea (E. Ostrom 2007; E. Ostrom et al. 2007; E. Ostrom and Cox 2010). The problems associated with monocentricity, for instance, do not evaporate because some bureaucrats are encouraged to act entrepreneurially. However, some of the worst aspects of bureaucratic responses to disasters can potentially be dulled if certain bureaucrats were given standing orders to do what they think is in the best interest of their communities when disasters occur.

Eliminate, Suspend, or Simplify the Rules That Hamper Post-Disaster Entrepreneurship

There are a number of policies that might make sense in mundane times but hamper post-disaster entrepreneurship and slow community recovery. This might include the requirements and waiting periods for constructing a building and starting a business, as well as environmental regulations and occupational licensing. This might also include a number of rules that govern various disaster assistance programs that exist.

Consider, for instance, the regulations concerning child-to-staff ratios and center director qualifications, as well as the rules governing physical structures of out-of-home child care centers. In the pre-disaster context, these rules are meant to ensure that day care centers are safe and to address concerns that having too few qualified adults working in a center will not be in the children's interest. After a disaster, however, these policies make it harder for child care centers to reopen and expand their services since many of the qualified employees they might employ and many of the buildings where they may set up shop are likely to have also been affected by the disaster. In this situation, child care centers may have to remain closed until they can comply with regulations, instead of being able to operate in a minimal capacity until they can make renovations and hire additional staff. As such, these policies make it difficult for parents to find adequate child care as they attempt to rebuild their homes and return to work. Not surprisingly, child care posed an acute problem in post-Katrina New Orleans. More than two years after the storm, for instance, only 94 of the 275 pre-storm day care centers had reopened, and several that did open were forced to operate below pre-Katrina levels because they were unable to employ enough staff (Gyan 2007; Bronston 2006).

Several of the entrepreneurs that we discuss in chapters 5, 6, and 7 had their efforts slowed by regulatory hurdles. Alice Craft-Kerney, for instance, faced unexpected delays in opening her clinic in the Lower Ninth Ward due to permit issues. Although she had a residential permit for the building where the clinic was housed, it was argued that her clinic violated residential zoning laws and that, instead, she needed a commercial permit to provide health care services to the public. Additionally, officials were concerned that the handicap ramp only had a hand railing on one side, which is in violation of building and safety codes for establishments like the clinic. These regulatory hurdles delayed the opening of the clinic by almost six months. Ironically, her clinic closed a few years later because of a change in regulatory requirements. Similarly, Casey Kasim was not permitted to reopen his laundromat for several weeks after it was completely rebuilt because he had to wait for the city's inspector to first inspect the facility and then forward the necessary paperwork to Entergy, the power company servicing the Lower Ninth Ward.

Additionally, relaxing regulations, like occupational licensing, can foster recovery without necessarily increasing the issues that such regulations attempt to deter. Skarbek (2008), for instance, found that relaxing occupational licensing requirements in Florida after Hurricanes Frances and Katrina allowed more roofers to enter the market and repair homes. Furthermore, there was little evidence of asymmetric information issues (where roofers had more information about their skills than home owners), which is the primary reason for adopting occupation-licensing requirements. This is a substantial finding, since the post-disaster context is filled with uncertainty, which would likely exacerbate any asymmetric information issues. Given these findings, Skarbek (ibid.) argues that such requirements should not simply be relaxed after a disaster but should be reformed permanently.

Furthermore, Walling (2006) highlighted how numerous petroleum, fuel, and gasoline regulations were relaxed at the state and federal levels of government following Katrina. First, the federal government waived regulations that allowed fuel companies to release stores of fuel typically held for the winter, and also relaxed restrictions on using high-sulfur diesel. Second, the federal government waived requirements related to the use of boutique fuel, which is "for use in markets that cannot meet federal air quality standards without specialty fuel" (Walling 2006: 10). Third, governments refrained from implementing price controls, which would have only exacerbated fuel shortages. This strategy proved effective at alleviating shortages.

Policymakers should consider eliminating, suspending, or simplifying the rules that hamper post-disaster entrepreneurship. Moreover, policymakers should be clear about which regulations will be enforced and which they intend to relax (and for how long) before a disaster occurs, since uncertainty about the regulatory environment hampers entrepreneurial activity and, consequently, community recovery. They can accomplish this by developing an alternative regulatory regime before a disaster actually occurs that would be implemented following a disaster. This alternative regime would relax regulations and requirements that hamper entrepreneurship after disasters.

This alternative regime would also reduce noncompliance, as entrepreneurs would have a clearer sense of which regulations government officials intend to enforce, which they are unwilling to change, and which they are willing to disregard. Additionally, this alternative regulatory regime would reduce delays that result when governments are forced to make ad hoc regulatory concessions after a disaster.

In Summary, Communities That Exhibit Polycentricity Are More Likely to Be Resilient

There is a significant collective action problem associated with post-disaster situations. While bureaucrats in monocentric systems are not well suited to overcome the problem, entrepreneurs in polycentric systems are well suited and do regularly find ways to overcome the problem and spur recovery. Based on this evidence, the appropriate role for policymakers is much more limited than currently assumed. Instead of focusing on leading and controlling disaster recovery efforts, policymakers should focus on reforms that create a space for entrepreneurs to act. Such reforms include ensuring that even bureaucrats have the freedom and incentive to act entrepreneurially to meet disaster victims' needs, as well as eliminating, suspending, or simplifying the rules that hamper post-disaster entrepreneurship.

Chapter 9

Conclusion

This book has advanced three main arguments, specifically that (1) entrepreneurship occurs across all sectors and entrepreneurs should be viewed as social change agents, (2) entrepreneurs are key drivers of community recovery after disasters, and (3) giving entrepreneurs the space to act is crucial for fostering resilient communities.

Entrepreneurs Are Agents of Social Change

Our notion of who counts as an entrepreneur is quite inclusive, encompassing commercial entrepreneurs as well as social, political, and ideological entrepreneurs. Think of the businessman who opens a grocery store in what had been considered a "food desert" or builds a fitness center. Or think of the community leader who starts a food pantry for the disadvantaged in her community or raises funds to build a recreational center. Or think of the social activist who promotes programs that encourage English as a second language and who provides legal assistance to immigrants. While there are differences in how entrepreneurs function in different environments, we argue that these differences are overstated and that their similarities are understated. Indeed, the commercial entrepreneur that opens a grocery store and the social entrepreneur that establishes a food pantry both provide food to community members. Both the social entrepreneur who builds a recreational center and the commercial entrepreneur who builds a fitness center offer a space where community members can exercise. Different types of entrepreneurs are often alert to similar opportunities, often provide similar goods and services, and often engage in similar activities.

Although their experiences do lead them to see certain opportunities that others do not, entrepreneurs are not able to notice these opportunities to bring about social change because they possess superior skill or expert knowledge. Instead, they are simply alert to and discover opportunities that others have overlooked. They notice opportunities other actors would have been able to see if only they were looking in that direction. As Kirzner (1994: 107) describes, alertness is a "generalized intentness upon noticing the useful opportunities that may be present within one's field of vision." Indeed, the entrepreneurs we discuss in chapters 5, 6, and 7 share a common alertness to opportunities to promote community recovery rather than any set of skills or personality traits. Recall, for instance, the experiences of Mary Ann Patrick, who ran an appliance and furniture store in Chalmette before Hurricane Katrina and lost everything in the storm. Like many residents of New Orleans, she was living in a trailer as she started her cleanup and rebuilding efforts. Patrick knew appliances and furniture, and like her customers, she was a victim of Hurricane Katrina's devastation. Exhausted from countless hours of cleaning, she understood that a bed to sleep in was important. Similarly, she understood that washers and dryers were high priority items for returning disaster victims. Patrick was alert to the interest in certain appliances and furniture and adjusted her inventory to suit the demands of her clients. In Bayswater on the Rockaway Peninsula, Rabbi Kruger also noticed that the residents in his community needed mattresses, water heaters, and refrigerators after Hurricane Sandy. As an active member in the community and a representative of the Community Assistance Fund (CAF), Rabbi Kruger was keenly aware of the diversity of needs. He decided to capture the specific needs of each member of the community in a spreadsheet and used that tool to connect aid demanders with suppliers of disaster assistance. Likewise, Father Vien, in his service to his parishioners of the Mary Queen of Vietnam Catholic Church, recognized what his community needed to return and recover and worked to satisfy those needs.

We have argued that whether the entrepreneur is acting in a market context and reading prices and profit or loss, or she is acting in a nonmarket context and reading other signals (such as reputation), she receives feedback on her actions. Rabbi Bender on the Rockaway Peninsula in New York, for instance, expanded the efforts of the Achiezer Community Resource Center following Hurricane Sandy and recognized that reinstating CAF would serve as a way to raise, manage, and distribute donations. The feedback Rabbi Bender and Achiezer received suggested that they properly identified needs in the community and was viewed as capable of meeting those needs. One signal was the increase in call volume. As Rabbi Bender indicated, Achiezer fielded approximately 1,500 phone calls a day in the week following Hurricane Sandy. Another piece of feedback was the millions of dollars donated to CAF. Presumably, if people did not believe it to be a worthy cause, or did not believe that Achiezer would be able to distribute the assistance to those who needed it, they would not have given so generously. Our interviews with community members illustrate that Achiezer has a strong reputation. Likewise, the hours that Father Vien spent calling and visiting displaced members of his congregation resulted in a large response from his community members. The first service that Father Vien held after Hurricane Katrina on October 9 attracted 300 parishioners. By the end of October, over 2,000 parishioners were in attendance. Father Vien interpreted the attendance as a sign that community members did indeed want to return and rebuild the MQVN community and that his efforts to encourage them to return and rebuild were working.

In the market context, the mechanism of profit and loss is understood to provide feedback to the entrepreneur. There is reason to believe, however, that feedback in the market context is not as unambiguous as it is sometimes portrayed. Imagine, for instance, that a baker opens a store and, at the end of the first month, completes her bookkeeping and sees that she has received a loss. Does she close the doors at that moment and release her capital to be used somewhere else where it may be more highly valued? Or does she decide to wait another month? Another six months? Another year? Moreover, how does she interpret the loss? Did she fail to earn a profit because customers liked her breads but not her pastries? Have pastry lovers not discovered her bakery? Is it an issue with her service or her products? Recognizing that even profit and loss signals need to be interpreted means admitting that the differences in the feedback mechanisms available in market and nonmarket contexts are not as stark as they sometimes appear to be.

Whether they are acting within markets or in nonmarket settings, entrepreneurs advance broader social change through their individual actions. For instance, Ed Williams of the Lower Ninth Ward repaired roofs in New Orleans and in doing so made it easier for residents of the community to return. Mrs. Russell started a meal service in Bayswater that provided not only nourishment for disaster victims but also a space where they could socialize, obtain information about disaster assistance, and develop plans for recovery.

Entrepreneurs Are Drivers of Post-Disaster Community Recovery

Entrepreneurs play an important role in helping community members overcome the collective action problem that characterizes post-disaster recovery by (1) providing needed goods and services, (2) restoring and replacing disrupted social networks, and (3) signaling that recovery is likely and is, in fact, under way. By performing these functions, entrepreneurs affect the calculation of disaster victims concerning whether or not to return and rebuild by increasing the probability that others will return, increasing the benefits of returning, and decreasing the costs of returning and rebuilding.

In chapter 5, we discussed how entrepreneurs notice and pursue opportunities to provide needed goods and services to disaster victims. For instance, Alice Craft-Kerney, a nurse from the Lower Ninth Ward, recognized that many people in her community faced health issues after Hurricane Katrina. To address this need, she opened a health clinic in the neighborhood. Similarly, Jason Shtundel believed that Bayswater would benefit from increased security and established the Rockaway Citizen Safety Patrol. After Hurricane Sandy, the patrol extended its hours and functions to meet the post-disaster needs of the community.

After disasters, entrepreneurs also work to restore and replace disrupted social networks. As we showed in chapter 6, social networks play an important role in community recovery. In Broadmoor, for instance, LaToya Cantrell utilized the preexisting structure of the Broadmoor Improvement

Association to contact displaced residents, collect data to prove the neighborhood's vitality, and encourage rebuilding. Similarly, Ben Cicek opened a coffee shop in Chalmette after Hurricane Katrina, providing a needed social space that helped to restore and even create relationships among residents in the area who were trying to navigate the recovery process. Likewise, in Gentilly, Sandra Green facilitated the reconstruction of social capital as she greeted customers in her women's clothing store and exchanged stories and emotional support. Entrepreneurs are the custodians of social spaces where social networks are created, restored, and sustained.

In chapter 7, we explored how entrepreneurs act as signals that recovery is likely and, in fact, under way. Father Vien, for example, through phone calls and personal conversations, encouraged community members to return to New Orleans East. At the same time, he worked to restore important club goods, including religious services and ethically appropriate aid. Additionally, Doris Voitier, superintendent of the St. Bernard Parish Unified School District, was aware that the decision of many displaced community members to return and rebuild was linked to the restoration of local schools. Parents did not want their children's education to be disrupted, and they needed a place for children to go during the day so that they could repair their homes and return to work. Voitier worked around the clock to reopen the schools in St. Bernard Parish. Father Vien and Voitier were successful in helping their communities overcome the collective action problem that characterizes post-disaster recovery, in part, because of their local knowledge, reputation, and social ties.

It Is, Thus, Important to Create Space for Entrepreneurs

Finally, in chapter 8, we argue that recovery is achieved through the entrepreneurial activities of community leaders and the decentralized actions of residents, and that fostering resilient communities requires ensuring entrepreneurs have the space to act. Environments that promote entrepreneurship are necessarily flexible and decentralized; they are polycentric systems. Polycentric systems, as opposed to monocentric systems, provide the space for entrepreneurs to drive social change. Policymakers should, thus, work to foster entrepreneurial activity. If policymakers focus on creating a space for entrepreneurs to act, they will foster not only post-disaster community recovery but also resilient communities that can withstand and quickly rebound from future disasters and crises.

Erratum to:
Chapter 3 in: *How Entrepreneurs Promote Post-Disaster Community Rebound,*

There are two errors in the formulas in this chapter and the error has been corrected. The errors are below:

- Second line on page 36. The formula $\beta > (\alpha - c)\ P_r + (\alpha - \delta)\ (1 - P_r)$ should be $\beta > (\alpha - c)\ P_r + (\delta - c)\ (1 - P_r)$
- First line on page 37. The formula $((\alpha + \theta - c + \gamma)\ P_r + (\alpha + \theta - \delta)\ (1 - P_r)) > \beta$ should be $(\alpha + \theta - c + \gamma)\ P_r + (\delta + \theta - c + \gamma)\ (1 - P_r) > \beta$

The updated original online version for this chapter can be found at
DOI 10.1007/978-1-137-31489-5_3

DOI 10.1007/978-1-137-31489-5_10

Notes

1 Introduction

1. See Fothergill and Peek ("Poverty and Disasters in the United States: A Review of Recent Sociological Findings," *Natural Hazards* 32, no. 1 (2004): 89–110) for a survey on the adverse effects of disasters related to wealth, income, and poverty in the United States.
2. This, of course, assumes that post-disaster recovery is desirable.
3. See chapter 3 for more details on the collective action problem inherent in post-disaster recovery as well as the strategies that disaster victims use as they go about recovery or decide to start anew elsewhere. By saying that deciding to "wait and see" could be the dominant strategy for all community members, we simply mean that since the expected benefits of waiting are likely to be higher than the expected costs associated with waiting, the sensible move for community members could be to let others decide to return or relocate before deciding what to do.
4. As Chamlee-Wright (*The Cultural and Political Economy of Recovery: Social Learning in a Post-Disaster Environment* (London: Routledge, 2010): 1–2) similarly describes,

   > A successful return required residents to solve simultaneously several problems, many of which were out of their immediate control. A returning resident needed a place to stay, a job, financial resources for rebuilding, schools for their children, transportation, and services of utilities, area businesses, and local government. Businesses additionally needed clients and employees. Absent some orchestrated effort, the residents and business owners that moved back first took on disproportionate risk. But if everyone waited for everyone else to move back first, the community would fail to rebound. In short, the post-Katrina context presented a collective action problem of significant proportion.

5. Key entrepreneurship studies scholars have also recommended this inclusive conception of entrepreneurship. Drucker ("Entrepreneurial Strategies," *California Management Review* 27, no. 2 (1985): 28), for instance, has described the entrepreneur as someone who "*always searches for change, responds to it, and exploits it as an opportunity.*" As he explains, changes in demographics, technological changes, and public perceptions create opportunities for entrepreneurial innovation. Although Drucker focuses primarily on commercial entrepreneurship, he acknowledges that entrepreneurship can occur in noncommercial spheres. As Drucker (1985: 27) writes, "entrepreneurship

is by no means limited to the economic sphere . . . the entrepreneur in education and the entrepreneur in health care . . . do very much the same things, use very much the same tools, and encounter very much the same problems as the entrepreneur in a business or a labor union."

6. Aberbach et al. (*Bureaucrats and Politicians in Western Democracies.* (Cambridge, MA: Harvard University Press, 1981)) have described this as a "determined centrism." As they (ibid.: 166) describe, "it might be tempting to conclude that bureaucrats are 'nonideological,' immune to the ideological imperative. But our evidence speaks against this view. The very consistency and coherence of the bureaucrat's centrism suggests that it *is* an ideology. The bureaucrats we interviewed are not randomly centrist, flip-flopping from issue to issue. They are determinedly centrist."
7. See, for instance, Birch and Wachter (eds., *Rebuilding Urban Places After Disaster: Lessons From Hurricane Katrina* (Philadelphia: University of Pennsylvania Press, 2006)); Schneider ("Who's to Blame? (Mis)perceptions of the Intergovernmental Response to Disasters," *Publius: The Journal of Federalism* 38, no. 4 (2008): 715–738); Thaler and Sunstein (*Nudge Improving Decisions about Health, Wealth and Happiness* (New York: Penguin Press, 2008)); Cigler ("Post-Katrina Hazard Mitigation on the Gulf Coast," *Public Organization Review* 9, no. 4 (2009): 325–341); Springer ("Emergency Managers as Change Agents," *Ideas from an Emerging Field: Teaching Emergency Management in Higher Education* 12, no. 1 (2009): 197–211; "Achieving Community Preparedness Post-Katrina," *UNLV Institute for Security Studies Faculty Publications* (2011)).
8. See, for instance, Boettke and Prychitko ("Is an Independent Nonprofit Sector Prone to Failure?" *The Philanthropic Enterprise* 1, no. 1 (2004): 1–63); Boettke and Coyne (*Context Matters: Institutions and Entrepreneurship* (Hanover: Now Publishers Inc., 2009a)); Coyne ("The Importance of Expectations in Economic Development," *The Wealth and Well-Being of Nations* 1 (2009): 63–82; *Doing Bad by Doing Good* (California: Stanford University Press, 2013)).

2 The Entrepreneur as a Driver of Social Change

1. Radical uncertainty, the notion that uncertainty is present in all action and all choices, is a main tenet of the Austrian school of economics (Kirzner, "The Austrian School of Economics," in *New Palgrave Dictionary of Economics*, ed. S. N. Durlauf and L. E. Blume (New York: Macmillan, 1986); Leeson and Boettke, "The Austrian School of Economics: 1950–2000," in *A Companion to the History of Economic Thought*, ed. Warren Samuels, Jeff Biddle, and John Davis (Oxford: Blackwell Publishing Ltd., 2003: 445–453). See, for instance, Lachmann (*Capital, Expectations and the Market Process* (Kansas City: Sheed and McMeel, 1977)); and O'Driscoll and Rizzo (*The Economics of Time and Ignorance* (Oxford: Basil Blackwell, 1985)).
2. As Kirzner (*Competition and Entrepreneurship* (Chicago: University of Chicago Press, 1973): 69) explains, "a state of market disequilibrium is characterized by wide-spread ignorance. Market participants are unaware of the real opportunities for beneficial exchange which are available to them in the market. The result of this state of ignorance is that countless opportunities are passed up." Stated differently, market participants lack perfect knowledge and are often unaware of what they do not know.

3. Stated another way,

   For each product, as well as for each resource, opportunities for mutually beneficial exchange among potential buyers and sellers are missed. The potential sellers are unaware that sufficiently eager buyers are waiting, who might make it worth their while to sell. Potential buyers are unaware that sufficiently eager sellers are waiting, who might make it attractive for them to buy. Resources are being used to produce products which consumers value less urgently, because producers (and potential producers) are not aware that these resources can produce more urgently needed products. Products are being produced with resources badly needed for other products, because producers are not aware that alternative, less critically needed resources can be used to achieve the same results (Kirzner 1973: 69).

4. Other economists have different views on the role of the entrepreneur. For instance, Knight (*Risk, Uncertainty and Profit* (New York: Hart, Schaffner, and Marx, 1921); "Profit and Entrepreneurial Functions," *The Journal of Economic History* 2 (1942): 126–132) stresses that entrepreneurs take on risks in order to receive profits. Lachmann ("An Austrian Stocktaking: Unsettled Questions and Tentative Answers," in *New Directions in Austrian Economics*, ed. Louis M. Spadaro (Kansas City: Sheed Andrews and McMee, 1978): 1–18; *The Market as an Economic Process* (Oxford: Basil Blackwell, 1986); and "Austrian Economics as a Hermeneutic Approach," in *Economics and Hermeneutics*, ed. Don Lavoie (London: Routledge, 1991): 134–146) recognizes that the entrepreneur is alert to opportunities within the context of the market process, but argues that there is no guarantee that the entrepreneur's actions are equilibrating.
5. As Kirzner (1973: 67) states, though "the element of knowledge is tied to the possibility of winning pure profits, the elusive notion of entrepreneurship is, as we have seen, not encapsulated in the mere possession of greater knowledge and market opportunities. The aspect of knowledge which *is* crucially relevant is not so much the substantive knowledge of market data as *alertness, the 'knowledge' of where to find market data.*" Further, Kirzner (*Perception, Opportunity, and Profit: Studies in the Theory of Entrepreneurship* (Chicago: University of Chicago Press, 1979): 8) states, "entrepreneurial knowledge is a rarefied, abstract type of knowledge—the knowledge of where to obtain information (or other resources) and how to deploy it."
6. Note also that to calculate the costs and benefits of a deliberate search, an individual must possess some degree of knowledge about the landscape and the likelihood of finding what he hopes to find. That knowledge of the landscape and the likelihood of finding what he is looking for must have necessarily been given to him in advance of his search.
7. As Kirzner (*Discovery and the Capitalist Process* (Chicago: University of Chicago Press, 1985): 21–22) states, "If an entrepreneur's discovery of a lucrative arbitrage opportunity galvanizes him into immediate action to capture the perceived gain, it will not do to describe the situation as one in which the entrepreneur has decided to use his alertness to capture this gain. He has not 'deployed' his hunch for a specific purpose; rather his hunch has propelled him to make his entrepreneurial purchase and sale. The entrepreneur never sees his hunches as potential inputs about which he must decide whether they are to be used."

8. See Klein and Bylund ("The Place of Austrian Economics in Contemporary Entrepreneurship Research," *The Review of Austrian Economics* 27, no. 3 (2014): 259–279) for a survey of some of the criticisms of Kirzner's theory of entrepreneurship.
9. Although this conception of development, conceived of as "the carrying out of new combinations," is not entirely incompatible with mainstream conceptions in economics, several salient features of Schumpeter's formulation distinguish his view from the ones held by the bulk of the discipline. One fundamental difference is the way in which capital and credit are employed. In standard formulations, economic development is principally a matter of capital accumulation; with capital treated as homogenous fully divisible units of some "thing" that enters into an aggregate production function and is transformed into output (See Jones, *Introduction to Economic Growth* (New York: W.W. Norton & Company, 1998); and the first three chapters of Romer (*Advanced Macroeconomics, 4th ed.* (New York: McGraw-Hill Irwin, 2012)) for an overview of the macroeconomic literature on economic growth). Although Schumpeter also emphasizes the importance of credit and capital, development in his scheme is not reduced to the outcome of some aggregate phenomenon. That society is adding to its capital stock is less important than how individuals are using its current stock and which individuals are directing its use. It matters for Schumpeter whether an *entrepreneur qua entrepreneur* or a businessman is controlling capital and deciding on how it is used. Another distinction, which is perhaps implicit in the difference in orientation expounded above, regards how standard formulations and Schumpeter treat growth. The macroeconomic literature talks about economic growth and economic development as if they are synonymous. By development, Schumpeter, however, has in mind something more fundamental.
10. Schumpeter (*The Theory of Economic Development: An Inquiry into Profits, Capital, Credit, Interest, and the Business Cycle* (New Brunswick: Transaction publishers, [1934] 2012): 65) continues: "not in the sphere of the wants of the consumers of final products." Schumpeter (ibid.: 65) also notes that,

    > We must always start from the satisfaction of wants, since they are the end of all production, and the given economic situation at any time must be understood from this aspect. Yet, innovations in the economic system do not as a rule take place in such a way that first new wants arise spontaneously in consumers and then the productive apparatus swings around to the pressure. We do not deny the presence of this nexus. It is, however, the producer who as a rule initiates economic change, and consumers are educated by him if necessary; they are, as it were, taught to want new things, or things which differ in some respect or other from those which they have been in the habit of using.

    It is on this basis that Schumpeter can claim that the entrepreneurial function is disequilibrating, a claim that Kirzner disagrees with. It is also on this basis that Schumpeter is able to deal with development from the perspective of "commercial life" and is therefore able to assert the importance of the entrepreneur in explaining the process.
11. Schumpeter ([1934] 2012: 66) also includes a fifth point: "The carrying out of the new organisation of any industry, like the creation of a monopoly position (for example through trustification) or the breaking up of a monopoly position."

12. Schumpeter ([1934] 2012: 70): "talent in economic life 'rides to success on its debts'...it is as clear *a priori* as it is established historically that credit is primarily necessary to new combinations and that it is from these that it forces its way into the circular flow, on the one hand because it was originally necessary to the founding of what are now the old firms, on the other hand because its mechanism, once in existence, also seizes old combinations."
13. As Schumpeter ([1934] 2012: 84–85) writes,

    "Carrying out a new plan and acting according to a customary one are things as different as making a road and walking along it." Also, as he (ibid.: 86) explains, "In the breast of one who wishes to do something new, the forces of habit rise up and bear witness against the embryonic project. A new and another kind of effort of will is therefore necessary in order to wrest, amidst the work and care of the daily round, scope and time for conceiving and working out the new combination and to bring oneself to look upon it as a real possibility and not merely as a day-dream. The mental freedom presupposes a great surplus force over the everyday demand and is something peculiar and by nature rare."

    Schumpeter also notes that social forces might also be opposed to and so react negatively against the introduction of something new. "This reaction," he (ibid.: 86–87) explains, "may manifest itself first of all in the existence of legal or political impediments. But neglecting this, any deviating conduct by a member of a social group is condemned, though in greatly varying degrees according as the social group is used to such conduct or not...In matters economic this resistance manifests itself first of all in the groups threatened by the innovation, then in the difficulty in finding the necessary cooperation, finally in the difficulty in winning over consumers."
14. Specifically, as Kirzner (1973: 31) writes, "there is present in all human action an element which...cannot itself be analyzed in terms of economizing, maximizing, or efficiency criteria." This position was also held by Mises (*Human Action: A Treatise on Economics* (New Haven: Yale University Press, [1949] 1963): 252), who stated that, "In any real and living economy, every actor is always an entrepreneur and speculator."
15. It should be noted that Schumpeter (*Capitalism, Socialism and Democracy* (London: George Allen & Unwin, [1942] 1976)) left room for political entrepreneurship within his concept of the entrepreneur. Specifically, he acknowledged that political leaders compete for power and control of government but did not further develop a notion of the political entrepreneur.
16. We do not here mean the "ideological entrepreneurship" discussed by Rose-Ackerman ("Altruism, Ideological Entrepreneurs and the Non-profit Firm," *Voluntas* 8 (1997): 120–134), who has in mind entrepreneurs who have certain ideological commitments that can act as a guarantee for donors that gifts will go toward the high quality provision of services.
17. For information on Operation Fresh Start, see: http://www.operationfreshstart.org.
18. Although they do not mention it, government failure in addition to market failure is surely also an opportunity for social entrepreneurs. It is also worth noting that social entrepreneurship may not need to solely act to fill a void created by other activities from other sectors. As mentioned above, civil society, including religion and charity, is often intertwined with and can serve as complements as well as substitutes for commercial and government activity.

19. The fact that it is difficult to determine and show success also means that social entrepreneurship can lead to negative as well as positive outcomes. Despite the best intentions, social endeavors can create waste and unintended outcomes rather than spur meaningful and positive social change (Boettke and Coyne, *Context Matters: Institutions and Entrepreneurship* (Hanover: Now Publishers Inc., 2009a); "An Entrepreneurial Theory of Social and Cultural Change," in *Markets and Civil Society: The European experience in Comparative Perspective* (2009b): *77–103*). However, there are clear examples of success, which suggests that the feedback mechanisms we discuss, such as reputation and direct monitoring, are effective guides for social entrepreneurship (Chamlee-Wright, "Local Knowledge and the Philanthropic Process: Comment on Boettke and Prychitko," *Conversations in Philanthropy* 1, no. 1. (2004): 45–51; Boettke and Coyne (2009a)).
20. There is also literature that discusses the limits of the bank as well as criticisms of the bank's use of subsidies and other practices. For instance, see Jain ("Managing Credit for the Rural Poor: Lessons from the Grameen Bank," *World Development* 24, no. 1 (1996): 79–89) and Morduch ("The Role of Subsidies in Microfinance: Evidence from the Grameen Bank," *Journal of Development Economics* 60, 1 (1999): 229–248).
21. There is also a growing literature on institutional entrepreneurship, where entrepreneurs seek to alter the informal and formal rules of society. For instance, see Eisenstadt ("Cultural Orientations, Institutional Entrepreneurs, and Social Change: Comparative Analysis of Traditional Civilizations," *American Journal of Sociology* 85, no. 4 (1980): 840–869); DiMaggio ("The New Institutionalisms: Avenues of Collaboration," *Journal of Institutional and Theoretical Economics* 154, no. 4 (1998): 696–705); Hwang and Powell ("Institutions and Entrepreneurship," in *Handbook of Entrepreneurship Research*, ed. S. A. Alvarez, R. Agarwal, and O. Sorenson (New York: Springer, 2005: 201–232); Boettke, Coyne, and Leeson ("Institutional Stickiness and the New Development Economics," *American Journal of Economics and Sociology* 67, no. 2 (2008): 331–358); Leca et al. (*Agency and Institutions: A Review of Institutional Entrepreneurship* (Cambridge: Harvard Business School Press, 2008)); and Boettke and Coyne (2009b). There is some literature on how this process may occur in the political arena. For instance, see Hall ("Policy Paradigms, Social Learning, and the State: The Case of Economic Policymaking in Britain," *Comparative Politics* 25, no. 3 (1993): 275–296) for an examination of how policies can change; and Hay ("Narrating Crisis: The Discursive Construction of the Winter of Discontent," *Sociology* 30, no. 2 (1996): 253–277) for how moments of crisis can lead to policy changes. Ideological entrepreneurship may be seen as a subset of institutional entrepreneurship, focusing on the informal rules grounded in beliefs and morals.
22. Recall that more than half the revenues nonprofits receive are through fees rather than charitable donations.
23. Furthermore, the motivations behind commercial, social, and ideological entrepreneurship are likely not as stark as often described. While commercial entrepreneurs are recognized as being able to bring about socially beneficial outcomes through self-interested profit-seeking behavior, social and ideological entrepreneurs are often characterized as needing more benevolent motivations to bring about positive social change. This is not necessarily the case. Indeed, social and ideological entrepreneurs may seek prestige, power, and/

or outcomes they personally prefer and, yet, their actions might still result in socially desirable outcomes.

24. It can, of course, be argued that social and ideological entrepreneurs need not pay attention to signals. That commercial entrepreneurs, on the other hand, cannot ignore profits and losses. Many business people, however, do in fact ignore profit and loss signals because they disagree with what the signals appear to be telling them, or because they have established their enterprises for a purpose other than maximizing profits. The commercial, social, and ideological entrepreneur who does not really desire social change would not be an entrepreneur in our view.
25. Martin ("Emergent Politics and the Power of Ideas," *Studies in Emergent Order* 3 (2010): 219) has likewise argued,

    Markets offer far tighter feedback mechanisms (regarding the extended order) than do polities. Agents in market environments are equipped with prices as ex ante signals to guide their conjectures and profits as ex post selection mechanisms to separate the wheat from the chaff. Combined with residual claimancy and transferable ownership, these constitute a tight feedback mechanism whereby individuals adjust their plans to conditions of scarcity determined on a global scale. Polities lack these key institutional features as well as any close substitutes. Agents launching enterprises in the political sphere thus stand in the default state of ignorance regarding the contours of the extended order.

    As Martin acknowledges, however, any assessment of the feedback mechanisms available in different spheres in the real world would depend on the (institutional) distance between actual markets and polities and their theoretical equivalents. Our analysis is not primarily concerned with the feedback mechanisms available to theoretical entrepreneurs but to actual entrepreneurs (in post-disaster contexts). Arguably, the (institutional) difference between actual and theoretical markets can be quite large or, at least, large enough to affect our assessments of their epistemic positions.
26. Stated another way, because they do not have access to profit and loss, entrepreneurs operating in non-priced environments cannot engage in rational economic calculation. According to Boettke and Prychitko ("Is an Independent Nonprofit Sector Prone to Failure?" *The Philanthropic Enterprise* 1, no. 1 (2004): 22), "although non-profits can undertake *measurements*, and, if encouraged, a rational assessment of their outcomes (using both quantitative and qualitative means), they have no way of calculating the realized results against the expected results. Nonprofit organizations and associations cannot, in other words, calculate the residual or monetary value-added of their endeavor, *ex ante* or *ex post*. In this sense, *nonprofit* is a better term for those that don't price their service or product. There is *no* calculated monetary profit." As we argue below, however, profit and loss can also be blunt signals in some circumstances.
27. That it is possible for erroneous but popular ideologies to persist would seem to lend credence to North's consistent emphasis on path dependence. But, if ideological entrepreneurs (along the lines developed above) do in fact exist, then they could provide a way out of the lock-in that North emphasizes. They are a potential source of ideological innovations and also a potential source of institutional change away from the current path. Moreover, ideological entrepreneurs are a source of both positive and negative social change.

28. Government sponsored programs, however, are not as responsive because they can continue to function long after community support has eroded, since their continued funding and staffing are not dependent on their ability to get local buy-in or the willingness of stakeholders to pay. Thus, political entrepreneurship is much less likely to have reliable feedback mechanisms, as compared to social and ideological entrepreneurship, and will, therefore, be more prone to result in negative unintended consequences.
29. Again, this is not to deny that entrepreneurs are in a privileged epistemic position over, say, experts within markets. As Mises (*Economic Calculation in the Socialist Commonwealth* (Ludwig von Mises Institute, [1920] 1990)) explains, central planners cannot engage in rational economic calculation and so cannot centrally plan an economy. And, as Skarbek ("Experts and Entrepreneurs," *Experts and Epistemic Monopolies* 17 (2012): 99) suggests, "Entrepreneurs have the privileged epistemic positions capable of accessing their own creativity and preference rankings as well as the requisite local knowledge of time and place." What we are denying, however, is the notion that entrepreneurs in market settings always and necessarily enjoy an epistemic advantage over entrepreneurs in nonmarket settings. Even if it were true that, on average, commercial entrepreneurs occupy a more privileged epistemic position than social entrepreneurs, this would not say anything about the epistemic status of any particular commercial or social entrepreneur.

3 How Entrepreneurs Promote Post-Disaster Community Rebound

1. Schumpeter (*Capitalism, Socialism and Democracy* (London: George Allen & Unwin, [1942] 1976): 132) has compared his entrepreneur to the warrior classes of the past.
2. A similar version of this model appears in Chamlee-Wright and Storr ("Club Goods and Post-Disaster Community Return," *Rationality and Society* 21, no. 4 (2009a): 429–458).
3. See Chamlee-Wright ("The Long Road Back: Signal Noise in the Post-Katrina Context," *The Independent Review* 12, no. 2 (2007): 235–259; "The Structure of Social Capital: An Austrian Perspective on Its Nature and Development," *Review of Political Economy* 20, no. 1 (2008): 41–58) for a review of how regime uncertainty can distort the return calculus of displaced residents. Also see chapter 8.
4. This can be viewed as the list price associated with returning in the immediate aftermath of the disaster. This does not take into account any subsidies or discounts that the player might obtain. Nor does it include any decreases in the costs associated with returning that might come about because the prices of goods and services decrease for whatever reason.
5. This can be viewed as the benefits of returning to a post-disaster community that is similar to the pre-disaster community in all the relevant ways. It does not include the value of any goods or services that were not available in the pre-disaster community that might be available in the post-disaster community.
6. It is worth noting that θ and γ can be either positive or negative amounts, based on whether the new goods, services, and subsidies aid or hinder recovery. We discuss ways in which policymakers have hindered recovery efforts in chapter 8.

7. See, for instance, Chamlee-Wright and Storr ("Expectations of Government's Response to Disaster," *Public Choice* 144, no. 1–2 (2010b): 253–274) for a discussion of disaster victims' expectations of how government should, and will, respond after a disaster. It is worth noting that disaster victims did not always expect the government to provide extensive relief and recovery assistance. Skarbek ("The Chicago Fire of 1871: A Bottom-up Approach to Disaster Relief," *Public Choice* 160, no. 1–2 (2014): 155–180), for instance, finds that victims of the Chicago Fire of 1871 did not expect, nor rely on, government assistance because, at the time, the role of the government was more limited and did not readily include disaster response. See, for instance, Beito (*From Mutual Aid to the Welfare State: Fraternal Societies and Social Services, 1890–1967* (Chapel Hill, NC: University of North Carolina Press, 2002)) and Nelson ("The Chicago Relief and Aid Society 1850–1874," *Journal of the Illinois State Historical Society (1908–1984)* 59, no. 1 (1966): 48–66) for an alternate view concerning the private-led response to the 1871 Chicago Fire.
8. FEMA is a US federal government agency tasked with helping communities to prepare for and recover from disasters.
9. See Richter and Grasman ("The transmission of sustainable harvesting norms when agents are conditionally cooperative." *Ecological Economics* 93 (2013): 202–209) for an interesting discussion of how norms structure social networks.
10. There are, of course, several studies that point to some of the negative aspects of social capital. See, especially, Portes ("Social Capital: Its Origins and Applications in Modern Sociology," *Annual Review of Sociology* 24 (1998): 1–24) and Putnam ("Bowling Alone: America's Declining Social Capital," *Journal of Democracy* 6, no. 1 (1995): 65–78) for an analysis of the negative consequences of social capital; and Wacquant ("Negative Social Capital: State Breakdown and Social Destitution in America's Urban Core," *Netherlands Journal of Housing and the Built Environment* 13, no. 1 (1998): 25–40) for a discussion of the erosion of "state social capital" as a cause of urban blight.
11. See Chamlee-Wright and Storr ("Social Capital, Lobbying and Community-based Interest Groups," *Public Choice* 149, no. 1–2 (2011a): 167–185) for a discussion of the potentially negative aspects of social capital that can emerge after a disaster. Specifically, they argue that during the recovery process community leaders and residents can become adept at petitioning for resources and navigating bureaucratic red tape during recovery, leading to an increasing investment in lobbying social capital. After recovery, lobbying social capital can be used to continue seeking government services and funds rather than fostering a robust community based on self-governance.
12. See chapter 7 for a more in-depth discussion of the MQVN Catholic Church and the efforts of its pastor, Father Vien, to inspire his parishioners to return and to signal to them that community rebound was likely.
13. Admittedly, there are limits to the capacity of bonding social capital to be a source of mutual assistance after a large-scale disaster because others in a disaster victim's social network are likely to also be affected by the disaster. As Fussell ("Help from Family, Friends, and Strangers During Hurricane Katrina: Finding the Limits of Social Networks," in *Displaced: Life in the Katrina Diaspora*, ed. L. Weber and L. Peek (Austin: University of Texas Press, 2012): 150–151) describes, "Faced with a disaster like Hurricane Katrina, people

invariably turn to close family and friends to assist them in the evacuation and recovery... However, the geographic concentration of social networks within New Orleans, particularly those of low-income residents, led to a common problem: everyone in the network was affected by the disaster."

14. It should be noted that even homogenous groups must utilize their connections outside the community (or weak ties) to successfully access the resources needed for recovery. As such, some of the issues complicating recovery for heterogeneous, loosely connected communities will also impact homogenous, tightly connected communities.
15. Of course, it is likely that there are more readily available substitutes for weaker ties. One cordial but reasonably anonymous neighbor, for instance, is very similar to any other cordial but reasonably anonymous neighbor. That said, a displaced disaster victim would have no way of knowing that families similar to his pre-disaster neighbors would move in, should his pre-disaster neighbor decide not to return.
16. See chapter 6 for a more in-depth discussion of the Broadmoor Improvement Association and its then president LaToya Cantrell.
17. Again, see more details on the MQVN community and Father Vien in chapter 7.
18. Chapters 5, 6, and 7 include discussions of the efforts of entrepreneurs in St. Bernard Parish after Hurricane Katrina.
19. As E. Ostrom (*Governing the Commons: The Evolution of Institutions For Collective Action* (Cambridge: Cambridge University Press, 1990): 30) explains, "The term 'common-pool resource' refers to a natural or man-made resource system that is sufficiently large as to make it costly (but not impossible) to exclude potential beneficiaries from obtaining benefits from use."
20. Singleton and Taylor ("Common Property, Collective Action and Community," *Journal of Theoretical Politics* 4, no. 3 (1992): 309–324) have argued that E. Ostrom obscures the importance of community in her analysis and identifies several characteristics of a community: (1) shared beliefs, including normative beliefs and preferences, beyond those constituting their collective action problem, (2) a stable set of members, (3) an expectation of continued interaction among those members, and (4) the relationships among the members being direct and multiplex.
21. For E. Ostrom's treatment of social capital, see E. Ostrom ("Constituting Social Capital and Collective Action," *Journal of Theoretical Politics* 6, no. 4 (1994): 527–562; "Collective Action and the Evolution of Social Norms," *Journal of Natural Resources Policy Research* 6, no. 4 (2014): 235–252).
22. One way that communities do this is by creating environments where community members are incentivized to work together and where they see community rebound as being in their common interests (as we highlight in chapter 7, entrepreneurial action can act as a focal point for recovery that can inspire and encourage such activities). Seabright ("Managing Local Commons: Theoretical Issues in Incentive Design," *Journal of Economic Perspectives* 7, no. 4 (1993): 113–134) has offered a useful framework for discussing the role of incentives in shaping the efforts of communities to solve coordination problems.
23. Admittedly, Jacobs' notion of what constitutes a successful city neighborhood is not so mono-faceted. After all, successful neighborhoods house a variety of activities. For instance, Jacobs (*The Death and Life of Great American Cities* (New York: Random House Inc., 1961)) emphasizes diversity as being a key

characteristic of success. However, Weicher ("A Test of Jane Jacob's Theory of Successful Neighborhoods," *Journal of Regional Science* 13, no. 1 (1973): 29–40) and others find that there is little empirical support for Jacobs' claims that diversity leads to successful neighborhoods in terms of crime, disease, and death.

24. According to Jacobs (1961), the functions of self-government that these city neighborhoods perform differ depending on the size of the neighborhood in question. The functions of the different-sized neighborhoods also overlap in complex ways. For instance, if they are functioning well, street neighborhoods are expected "to weave webs of public surveillance and thus to protect strangers as well as themselves; to grow networks of small-scale, everyday public life and thus of trust and social control; and to help assimilate children into reasonably responsible and tolerant city life" (ibid.: 119). They also perform another more vital function, "they must draw effectively on help when trouble comes along that is too big for the street to handle" (ibid.: 119).

    Effective street neighborhoods must not only handle those issues that can be handled at the street level but must offer ways for neighbors to petition larger units for assistance when issues arise. Good district neighborhoods help to organize street neighborhoods to combat district-wide problems. Good district neighborhoods also communicate the needs of street neighborhoods to the city and secure resources for street neighborhoods from them. Districts, thus, act as intermediaries between the city as a whole and street neighborhoods. These need not, however, be contiguous with political units (that is, they need not be the same size as political wards or boroughs), but they have "to be large enough to count as a force in the life of the city as a whole... [and] big and powerful enough to fight city hall" (ibid.: 122). Effective district neighborhoods "help bring the resources of a city down to where they are needed by street neighborhoods, and they have to help translate the experiences of real life, in street neighborhoods, into policies and purposes of their city as a whole" (ibid: 122). Additionally, effective city districts "help maintain an area that is usable in a civilized way, not only for its own residents but for other users—workers, customers, visitors—from the city as a whole" (ibid: 122).
25. Many of the entrepreneurs we describe in this book (see chapters 5, 6, and 7) can be characterized as hop-skip people.
26. See chapter 8 for a more in-depth discussion of this point.
27. Schelling (*The Strategy of Conflict* (Cambridge: Harvard University Press, 1960)) has noted that a player can benefit from making a *strategic move* when involved in an interactive game.
28. See chapter 7 for more details on the entrepreneurial efforts of Casey Kasim and Father Vien.

4 How Entrepreneurship Promotes Community Recovery: The Cases of Hurricanes Katrina and Sandy

1. Available online at http://www.emdat.be/natural-disasters-trends. For an event to be considered a disaster, CRED specifies that "at least one of the following criteria must be fulfilled: Ten (10) or more people reported killed. Hundred (100) or more people reported affected. Declaration of a state of emergency. Call for international assistance." For more details, see http://www.emdat.be/explanatory-notes.

2. We should remember that every hurricane or earthquake is not a disaster in the sense we employ the term here. For us, these natural occurrences only "become" disasters if and when, as discussed in chapter 3, they cause large enough levels of destruction and affect and displace large enough numbers of community members such that responding and recovering from the occurrence constitute a collective action problem. Disasters, in this sense, are constructions. Hay ("Narrating Crisis: The Discursive Construction of the Winter of Discontent," *Sociology* 30, no. 2 (1996): 253–277) has suggested that crises are, similarly, social constructions.
3. These figures are real figures, with a base year of 1967.
4. CRED defines the number of people affected by disaster to be "people requiring immediate assistance during a period of emergency, i.e., requiring basic survival needs such as food, water, shelter, sanitation and immediate medical assistance." See the CRED website and explanatory notes available at http://www.emdat.be/explanatory-notes.
5. A large part of this occurrence can be attributed to increasing wealth, which translates to safer buildings and more advanced warning systems. Kahn ("The Death Toll from Natural Disasters: The Role of Income, Geography, and Institutions," *Review of Economics and Statistics* 87, no. 2 (2005): 271–284) performs a cross-country analysis and considers the impact of increasing incomes on the number of fatalities following disaster. He notes (ibid.: 280), "The average nation with a GDP per capita of $2000 experiences 944 deaths from natural disaster per year. If this nation's GNP per capita grew to $14,000, its death toll would fall to 180 per year."
6. FEMA data begins in 1974. There were 38 major declarations in 1975, 38 in 1990, and 45 in 2000.
7. Available online at https://www.fema.gov/disasters/grid/year/2013.
8. For example, Garrett and Sobel ("The Political Economy of FEMA Disaster Payments," *Economic Inquiry* 41, no. 3 (2003): 496–509) have shown that disaster assistance is not necessarily distributed based on damage. The authors locate two avenues for political influence to impact federal disaster assistance. First, the Stafford Act gives the president power to declare an emergency following a natural disaster, and because there are no strict criteria of emergency, the president may use personal discretion. Second, states represented on FEMA oversight committees may receive more disaster assistance. The authors examined disaster declarations and disaster assistance in the United States from 1991 to 1999 and found that states politically important to the president and states with congressional representatives on the FEMA oversight committees are more likely to have disaster declarations and receive more aid.
9. Over 1,000 lives were lost in the flooding in Uttarakhand, India. Further, the destruction and property damage in Uttarakhand threatened the livelihood of thousands more.
10. This data is from The Data Center (formerly the Greater New Orleans Community Data Center), available online at http://www.datacenterresearch.org/data-resources/katrina/facts-for-impact/.
11. A storm surge is caused by a few differing factors, including cyclonic winds and low pressure. Storm surges can cause extensive flooding and can affect areas tens of miles from the coast. According to the National Hurricane Center, storm surges are the greatest threat to life and property during a hurricane. The information about these two surges comes from testimony provided by

Peter Nicholson ("Hurricane Katrina: Why Did the Levees Fail?" United States Senate, One Hundred Ninth Congress, First Session, November 2 2005, http://www.gpo.gov/fdsys/pkg/CHRG-109shrg24446/html/CHRG-109shrg24446.htm.), on behalf of the American Society of Civil Engineers.

12. Information available online at http://www.uscg.mil/history/katrina/katrina index.asp.
13. Data available online at http://www.datacenterresearch.org/data-resources/katrina/facts-for-impact/.
14. Information available online at http://www.redcross.org/what-we-do/disaster-relief/hurricane-recovery-program.
15. Data available online at http://www.datacenterresearch.org/data-resources/katrina/facts-for-impact/.
16. Report available online at http://www.cdc.gov/mmwr/preview/mmwrhtml/mm5502a6.htm.
17. As part of the research post-Hurricane Katrina, we conducted 103 surveys and/or recorded interviews of New Orleans evacuees still living in Houston three years after the storm. Some preferred Houston to New Orleans; others preferred New Orleans to Houston. Chamlee-Wright and Storr ("There's no Place Like New Orleans: Sense of Place and Community Recovery in the Ninth Ward after Hurricane Katrina," *Journal of Urban Affairs* 31, no. 5 (2009b): 615–634) find that for those who prefer Houston to New Orleans, 35 percent indicated the reason was better schools, 35 percent found a better job in Houston, 33 percent indicated an overall higher quality of life, 31 percent noted lower crime, 27 percent reported better housing, and 21 percent indicated better access to health care. For those who preferred New Orleans to Houston (but nonetheless were located in Houston), 69 percent explained that New Orleans was home, 49 percent described the City's unique culture, 35 percent referred to an overall higher quality of life, 27 percent noted family/friend networks, 24 percent indicated better transportation, and 18 percent said that people were friendlier in New Orleans.
18. Data available online at http://www.nhc.noaa.gov/data/tcr/AL182012_Sandy.pdf.
19. This figure is reported by FEMA, available online at https://www.fema.gov/sandy-recovery-office.
20. Available online at https://www.fema.gov/news-release/2013/10/25/year-after-hurricane-sandy-new-jersey-recovery-numbers.
21. Available online at http://www.nysenate.gov/report/senate-bipartisan-task-force-hurricane-sandy-report.
22. The Federal Transit Administration's (FTA) Public Transportation Emergency Relief Program received $5.4 billion to repair transit systems affected by Hurricane Sandy. To repair damaged subway tunnels, $886 million was provided to the Metropolitan Transportation Authority.
23. Available online at http://www.state.nj.us/dep/special/hurricane-sandy/docs/rebuilding-after-sandy-factsheet.pdf.
24. With tourism as the third largest industry in New Jersey, business owners and politicians were eager to start rebuilding. Governor Chris Christie pledged tens of millions of federal Sandy relief money to the recovery in Seaside Heights. Information available online at http://www.nj.com/ocean/index.ssf/2014/09/a_year_after_devastating_boardwalk_fire_seaside_looking_to_start_new.html.

25. New York City is the largest city in the United States and has population increases each year. From 2012 (the year that Hurricane Sandy hit) to 2013, there were population increases in all boroughs. In New York City there was an increase of 0.83 percent. In the five borough populations, increases were in Bronx, 0.73 percent; in Brooklyn, 1.03 percent; in Manhattan, 0.44 percent; in Queens, 1.03 percent; and in Staten Island, 0.40 percent.
26. For more information, see http://www.nyc.gov/html/recovery/html/home/home.shtml.
27. Statistics found online here at http://www1.nyc.gov/office-of-the-mayor/news/498-14/mayor-de-blasio-marks-second-anniversary-sandy-major-progress-city-s-recovery-and#/0.
28. Qualitative methods are particularly useful in studying the response of individuals to disasters. As Vollmer (*The Sociology of Disruption, Disaster and Social Change: Punctuated Cooperation* (Cambridge: Cambridge University Press, 2013): 2) describes, "with increasing magnitude and relevance of a disruption, there is an intuitive sense that something is different after a disruption has taken place. In attempts to localize this difference, the impact on individuals, their biographies, their sense of order or psychological well-being tends to come into focus."
29. Additionally, as Chamlee-Wright (*The Cultural and Political Economy of Recovery: Social Learning in a Post-Disaster Environment* (London: Routledge, 2010): 27) explains, "to gain…perspective, we need to come down from time to time and look at our puzzle up close. When we do this, we need to admit that we are looking at a particular piece of the overall structure (not the whole thing) but this is true even when we look at it from the air. Though this on-the-ground view is partial, it affords us the opportunity to see details that would otherwise be missed."
30. 2010 Census data is available online at http://factfinder2.census.gov/faces/nav/jsf/pages/community_facts.xhtml.
31. Moreover, characterizing something as an entrepreneurial failure is not without its issues, since the "failures" teach the entrepreneur and others important lessons that then inform and inspire future entrepreneurial successes. Calling some entrepreneurial effort a failure is possibly demarcating what counts as the entrepreneurial act too narrowly and is possibly stopping the story prematurely. That said, we would not have highlighted entrepreneurial failures had they been easy to recognize.

5 Goods and Services Providers

1. Sometimes entrepreneurs are incentivized to provide necessary goods and services to disaster victims because the increased demand for certain goods and services post-disaster can result in their being sold at a higher price than is typical during mundane times. This, however, has been described as price gouging and has been viewed by some as being morally impermissible. We do not take a position on the moral permissibility or impermissibility of charging disaster victims a higher price after the disaster than they could have paid for a purchase before the disaster. See Zwolinski ("The Ethics of Price Gouging," *Business Ethics Quarterly* 18, no. 3 (2008): 347–378), however, for an ethical defense of price gouging.
2. For example, entrepreneurs organized Common Ground Relief and report that they provided the opportunity for 47,000 individuals to help in the

clean-up and rebuilding efforts in New Orleans following Hurricane Katrina. Similarly, Becky Zaheri, the founder of Katrina Krewe, had as many as 25,000 volunteers in New Orleans picking up trash and debris after the storm. For more information, see http://www.commongroundrelief.org/ and http://www.tolerance.org/supplement/katrina-krewe.

3. Whenever possible, we use pseudonyms, denoted with the symbol †, for the disaster victims we interviewed.
4. Data, based on the 2000 US census, is available online at http://www.gnocdc.org/prekatrinasite.html.
5. Data, based on the 2000 US census, is available online at http://www.gnocdc.org/prekatrinasite.html.
6. Report found online at https://gnocdc.s3.amazonaws.com/reports/Katrina_Rita_Wilma_Damage_2_12_06___revised.pdf.
7. Patrick's furniture store is still operational at the time of writing this book.
8. Data, based on the 2000 US census, is available online at http://www.gnocdc.org/prekatrinasite.html.
9. Also see http://www.pbs.org/newshour/bb/nation-july-dec13-lower9th_12-28/.
10. The Road Home Program is a federally funded disaster relief program for residents of Louisiana. Disaster funds were made available to individuals to repair and replace damaged homes and rental properties following Hurricanes Katrina and Rita. To date, over 130,000 residents received money through the program. Find more information online at https://www.road2la.org/.
11. Following Hurricane Katrina, Common Ground volunteers gutted over 3,000 homes. Currently, the organization provides community gardening and job training programs. For more information on the organization, see http://www.commongroundrelief.org/about-us.
12. See http://blog.nola.com/politics/print.html?entry=/2010/12/lack_of_finances_may_force_low.html.
13. See Kessler et al. ("Trends in Mental Illness and Suicidality after Hurricane Katrina," *Molecular Psychiatry* 13 (2008): 374–384) and Jaycox et al. ("Children's Mental Health Following Hurricane Katrina: A Field Trial of Trauma-Focused Psychotherapies," *Journal of Traumatic Stress* 23, no. 2 (2010): 223–231) for a discussion of the mental and general health issues that affected Katrina victims.
14. See https://www.nola.gov/nola/media/Health-Department/Publications/Health-Disparities-in-New-Orleans-Community-Health-Data-Profile-final.pdf.
15. Available online at http://kaiserfamilyfoundation.files.wordpress.com/2013/01/7659.pdf.
16. There is now a new health clinic in the Lower Ninth Ward, Baptist Community Health Services, Inc. For more information, see http://bchsnola.org/about-us/.
17. From the 2009–2013 American Community Survey 5-Year Estimates; individuals can select more than one race.
18. See http://www.rockawave.com/news/2013-11-29/Community/Rockaway_Citizens_Safety_Patrol_Reports.html.
19. There was reported looting in Belle Harbor, Coney Island, parts of Brooklyn, and Staten Island. Following Hurricane Katrina, there were many reports of looting. However, the literature suggests that the impression there was

rampant looting in New Orleans following Hurricane Katrina is fueled by sensational media reports rather than actual figures (Tierney et al., "Metaphors Matter: Disaster Myths, Media Frames, and Their Consequences in Hurricane Katrina," *The Annals of the American Academy of Political and Social Science* 604, no. 1 (2006): 57–81; Barsky et al., "Disaster Realities in the Aftermath of Hurricane Katrina: Revisiting the Looting Myth," *Natural Hazards Center Quick Response Report*, No. 184 (2006), http://udspace.udel.edu/bitstream/handle/19716/2367/Misc%20Report%2053.pdf?sequence=1.; Rodriguez et al., "Rising to the Challenges of a Catastrophe: The Emergent and Prosocial Behavior Following Hurricane Katrina," *The Annals of the American Academy of Political and Social Science* 604, no. 1 (2006): 82–101).

20. For more information, visit www.achiezer.org.
21. Testimonials available online at http://www.achiezer.org/testimonials.php.
22. There was an on-call emergency phone system in place (on a much smaller scale) long before Hurricane Sandy.
23. For more information, see this video on CAF: http://youtube/DuVIA6iJ3lQ.
24. The representatives served many functions during the process, often listening to people's stories and hardships and providing emotional support, as well as helping with applications and finding contractors and resources for rebuilding. Admittedly, some of these functions are part of their pastoral responsibilities in the community.
25. Rabbi Kruger's post-Sandy efforts are discussed in more detail in chapter 6.
26. Interview with Rabbi Cohen†.
27. Similar efforts to rebuild social networks are discussed further in the next chapter.

6 Regrowing Uprooted Social Networks

1. See Storr ("The Market as a Social Space: On the Meaningful Extra-Economic Conversations That Can Occur in Markets," *Review of Austrian Economics* 21, no. 2–3 (2008): 135–150) for a more in-depth examination of how commercial interactions can turn into tighter social bonds. See also Chamlee-Wright and Storr ("Commercial Relationships and Spaces after Disaster," *Society* 51, no. 6 (2014): 656–664) for the social bonds of commercial endeavors in the post-disaster context.
2. This ad hoc evacuation heightened the collective action problem that disaster victims had to overcome to bring about community rebound. As Weber and Peek (eds., *Displaced: Life in the Katrina Diaspora* (Austin: University of Texas Press, 2012): 11) continue, "these separations not only were emotionally traumatic, but also caused severe material and financial stress among families already living at the margins of poverty. Indeed, more often than not, the families separated by government relocation efforts had previously survived by pooling resources generated through extensive kin networks. Breaking up these networks left these evacuees with few alternatives for reuniting with their families or returning to the Gulf Coast."
3. Interview with Rabbi Cohen†.
4. For more information on the BIA, see http://www.broadmoorimprovement.com/.
5. For these statistics and more demographic data on Broadmoor, see The Data Center, available online at http://www.datacenterresearch.org/data-resources/neighborhood-data/district-3/Broadmoor/.

6. The organization, originally the Broadmoor Civil Improvement Association, is now known as the BIA. See the BIA's website for more information about the organization: http://www.broadmoorimprovement.com/#!news/c1ne.
7. See http://www.bestofneworleans.com/gambit/40-under-40/Content?oid=1246675.
8. For a more detailed bio on councilmember Cantrell, see http://nolacitycouncil.com/meet/meet_cantrell_long.asp.
9. Interview with Arnold O'Brien†.
10. For more information on BNOB, see http://www.bringneworleansback.org/.
11. See Cantrell ("Don't Sacrifice Our Homes to Flooding," *New Orleans Times Picayune*, last modified March 30, 2006, https://repository.library.brown.edu/studio/item/bdr:65484/); Abramson ("For One New Orleans School, an Uncertain Future," *National Public Radio*, last modified May 17, 2007, http://www.npr.org/templates/story/story.php?storyId=10228892); and "Life in the New Normal" (*Delta Sky*, last modified January 2008).
12. The Clinton Global Initiative and the Carnegie Corporation of New York donated $5 million in 2007 for the rebuilding of the Keller Library and other community foundations. The Surdna Foundation provided a total of $175,000 in 2007 and 2008 to support the efforts of the Broadmoor Development Corporation to assist home owners in rebuilding (Chamlee-Wright and Storr, "Filling the Civil Society Vacuum: Post Disaster Policy and Community Response," *Mercatus Center Policy*, Policy Comment No. 2 (Arlington, VA: Mercatus Center at George Mason University, February 2009c)).
13. For more information on their collaboration, see http://belfercenter.ksg.harvard.edu/project/54/broadmoor_project.html.
14. For more information, see http://nolacitycouncil.com/meet/meet_cantrell_long.asp.
15. For more details, see the Mercatus Center profiles on Ben Cicek (http://mercatus.org/sites/default/files/12_casestudy3Ben.pdf and https://www.youtube.com/watch?v=ujtj-BM1LrA).
16. Interview with Kevin Madison†.
17. Da Parish Coffee House is no longer open in Chalmette. Cicek, however, did successfully provide a needed good as well as a social space during the recovery process after Hurricane Katrina.
18. According to Chamlee-Wright (*The Cultural and Political Economy of Recovery: Social Learning in a Post-Disaster Environment* (London: Routledge, 2010): 122), "Gentilly respondents, for example, expressed appreciation for the distinct New Orleans music and cuisine, but given the more distinctly 'middle class' feel of the area and the more contemporary architectural styles, residents of this predominately middle class African-American neighborhood were much less likely than their Ninth Ward counterparts to emphasize the importance of socializing on the porch, block parties, and barbeques."
19. For these statistics and more, see http://www.datacenterresearch.org/pre-katrina/orleans/6/index.html.
20. As noted earlier, one of the particular challenges to recovery following a disaster, of the size and scope of Katrina, is that members of a disaster victim's social network are typically also affected by the disaster. As Fussell ("Help from Family, Friends, and Strangers During Hurricane Katrina: Finding the Limits of Social Networks," in *Displaced: Life in the Katrina Diaspora*, ed. L. Weber and L. Peek (Austin: University of Texas Press, 2012): 150) writes,

"In New Orleans members of these more vulnerable networks quickly reached the limits of the networks' ability to provide assistance precisely because of this concentration of similarly vulnerable people."

21. The clothing store is still operational at the time of writing this book.
22. For more demographic information, see http://www.city-data.com/neighborhood/Bayswater-Far-Rockaway-NY.html.
23. Interview with Tom Schmitz†.
24. The Bayswater Neighbors Fund was not designed to deal with widespread need and, thus, could not fill the needs of the neighborhood after Hurricane Sandy. Instead, they relied on CAF for storm-related assistance. Kruger continues to collect, maintain, and distribute resources for the Bayswater Neighborhood Fund.
25. For more information on Agudas Yisroel of Bayswater, see https://www.youtube.com/watch?v=fLcJi_78KVk.

7 Entrepreneurs as Signals of Healthy Community Rebound

1. Chamlee-Wright (*The Cultural and Political Economy of Recovery: Social Learning in a Post-Disaster Environment* (London: Routledge, 2010): 49) has described this as "non-priced signal effects of commercial network activity." As she explains, "to the extent that cooperation within commercial networks helps key service providers to return quickly, such activity helps to send a signal that the business community is committed to the rebuilding process, thereby aligning expectations of those still waiting to return."
2. According to the 1990 Census, this community was 87 percent Catholic (Bankston and Zhou, "De Facto Congregationalism and Socioeconomic Mobility in Laotian and Vietnamese Immigrant Communities: A Study of Religious Institutions and Economic Change," *Review of Religious Research* 41, no. 4 (2000): 453–470).
3. The initial wave of emigrants received help from Catholic Charities. For more information on the history of this community, see Min Zhou and Carl Bankston (*Growing Up American: How Vietnamese Children Adapt to Life in the United States* (New York: Russell Sage Foundation, 1998)).
4. According to Census 2000 Summary File 3 (SF-3), home ownership rates rose from 2.8 percent in 1980 to 27.8 percent in 1990, to 39.3 percent in 2000. In 1990, the median income in the MQVN neighborhood, Orleans Parish, and Louisiana were $17,044, $18,477, and $21,949, respectively. By 2000, median incomes were $24,955, $27,133, and $32,566, respectively.
5. See www.mqvncdc.org for information about the organization.
6. See Dunne ("Failure to Halt Landfill Doesn't Stop Activists," *The Advocate [Baton Rouge]*, last modified April 28, 2006) and Russell ("Chef Menteur Landfill Testing Called a Farce," New Orleans Times-Picayune, May 26, 2006, testimony of Reverend Nguyen, interview with James Bui, October 25, 2010).
7. For a more in-depth discussion on how pastors are enabled and constrained by their spiritual positions in the community after disasters, see Chamlee-Wright ("Pastor Response in Post-Katrina New Orleans: Navigating the Cultural Economic Landscape." In *Culture and Economic Action*, ed. L. E. Grube and V. H. Storr (Northampton: Edward Elgar 2015): 269–294).
8. Club goods are goods that are non-rivalrous (i.e., multiple people can consume the good without negatively impacting each other's consumption) but excludable (i.e., it is possible to prevent people who do not pay for the good

from consuming it). Stated another way, club goods are goods where the marginal cost of producing another unit is zero but where it is still possible to charge a positive price. Think, for instance, of private parks or satellite television.

9. For more demographic information on St. Bernard Parish, see http://www.gnocdc.org/prekatrinasite.html.
10. Interview with Pastor Brian†.
11. Cindy Mitchell† in an interview with Cindy Mitchell† and William Turner†.
12. William Turner† in an interview with Cindy Mitchell† and William Turner†.
13. For more details, see the Mercatus Center profiles on Casey Kasim (http://mercatus.org/sites/default/files/9_casestudy2caseykasim.pdf and https://www.youtube.com/watch?v=WIBeGpMLjRY).
14. Kasim knew his clientele well, stating that roughly 80 percent were local and 20 percent were traffic from the highway.
15. The report is available online at http://www.nolaplans.com/plans/Lambert%20Intermediate/District_8_Plan_FINAL%20PLAN%20REPORT%20Lower%20Ninth%20Ward-10-03-06.pdf.
16. Interview with Linda Rawls†.
17. Interview with Gerald Brandt†.
18. See Chamlee-Wright ("The Long Road Back: Signal Noise in the Post-Katrina Context," *The Independent Review* 12, no. 2 (2007): 235–259) for a discussion of how "signal noise" add to the uncertainty that disaster victims face after a disaster. Also see chapter 8.
19. Kasim's Discount Zone in the Lower Ninth Ward is no longer in operation. Kasim did, however, succeed prior to and after Katrina in creating a store that served as an anchor to the community.
20. Interview with Pastor Daniels†.
21. Interview with Ronda Miles†.
22. Interviews with Hester Chapman† and Trisha Thomas†.
23. The grocery store is still in operation at the time of writing this book.
24. For more demographic information on Far Rockaway, see http://www.city-data.com/neighborhood/Far-Rockaway-Far-Rockaway-NY.html.
25. Find out more here http://www.metcouncil.org/site/News2?page=NewsArticle&id=7941&printer_friendly=1.

8 Fostering Resilient Communities

1. Indeed, our analysis is silent on the question of whether public or private responses to disaster recovery are to be preferred. Rather, we argue that decentralized rather than centralized responses to disaster are to be preferred (i.e., polycentricity over monocentricity).
2. There have been several studies contrasting the effectiveness of monocentric versus polycentric orders in solving community-level problems. See, for instance, McGinnis (*Polycentricity and Local Public Economies: Readings from the Workshop in Political Theory and Policy Analysis* (Ann Arbor: University of Michigan Press, 1999b)) on domestic policing in the United States; Olowu and Wunsch (*Local Governance in Africa: The Challenges of Democratic Decentralization* (Boulder: Lynne Rienner Publishers, 2004)) on governance in Africa; Sproule-Jones (*Governments at Work: Canadian Parliamentary Federalism and its Public Policy Effects* (Toronto: University of Toronto Press,

1993)) on Canadian federalism; Sabetti (*Political Authority in a Sicilan Village* (New Brunswick: Rutgers University Press, 1984)) on governance in Italy; and Bogason ("The Fragmentation of Local Government in Scandinavia," *European Journal of Political Research* 30, no.1 (1996): 65–86) on governance in Scandinavia.

3. Monocentric systems also reduce the likelihood that the potentially diverse preferences of community members are catered to. As V. Ostrom and E. Ostrom ("Public Goods and Public Choices," in *Polycentricity and Local Public Economies. Readings from the Workshop in Political Theory and Policy Analysis*, ed. M. McGinnis (Ann Arbor: University of Michigan Press, [1977] 1999): 79–80) note, "Where a good is characterized by jointness of consumption and nonexclusion, a user is generally unable to exercise an option and has little choice whether or not to consume... Furthermore, individuals may be forced to consume public goods that have a negative value for them... Yet, the structure of institutional arrangements may have some effect on the degree of choice that individuals have. Councilmen representing local wards would, for example, be more sensitive to the protests by local residents about how streets are used in those wards than councilmen elected at large."
4. Of course, not all local officials can be entrepreneurs, neither can all managers of business or nonprofit organizations. For example, Sandra Green was highlighted as an entrepreneur not because of her title as general manager of a clothing store but because she turned her store into a social space for customers, where they could visit with one another and share their stories and frustrations.
5. Importantly, the way disaster victims perceive government's ability and intent to provide disaster assistance influences their decisions and actions after a disaster. Chamlee-Wright and Storr ("Expectations of Government's Response to Disaster," *Public Choice* 144, no. 1–2 (2010b): 253–274) find that when citizens are optimistic about the government's capability but pessimistic about its intent to provide relief, they will pursue a mixed strategy that includes self-directed recovery as well as appealing to the government for resources and reforms. However, this mixed strategy, which includes investing time and effort in engaging with the political process, can lead to "meeting fatigue" or "Katrina burnout." Such activities can slow recovery, since each day spent engaging in the political process means less time spent actually rebuilding.
6. As noted earlier, for an examination of the ethics of price gouging, and the moral argument for letting prices adjust in times of crisis, see Zwolinski ("The Ethics of Price Gouging," *Business Ethics Quarterly* 18, no. 3 (2008): 347–378).
7. Rodin (*The Resilience Dividend: Being Strong in a World where Things Go Wrong* (New York: Public Affairs, 2014)) created a framework based on successful efforts by communities, organizations, and individuals that focuses on awareness, adaptability, diversity, integration, and self-regulation. She emphasizes the importance of leadership and entrepreneurship for community resilience.
8. Paton et al. ("Disaster Response: Risk, Vulnerability and Resilience," *Disaster Prevention and Management* 9, no. 3 (2000): 173–179) and Paton ("Disaster Resilience: Building Capacity to Co-exist with Natural Hazards and Their Consequences," in *Disaster Resilience: An Integrated Approach*, ed. D. Paton and D. Johnston (Springfield: Charles C Thomas Publisher, 2006): 3–10)

argue that there are four general characteristics of resilient communities. Resilient communities, Paton (2006: 9) argues, must possess the following:

> Firstly, communities, their members, businesses and societal institutions must possess the resources (e.g. household emergency plans, business continuity plans) required to ensure, as far as possible, their safety and continuity of core functions in a context defined by hazard consequences (e.g. ground shaking, volcanic ash fall, flood inundation) that can disrupt societal functions. Secondly, they must possess the competencies (e.g. self-efficacy, community competence, trained staff, disaster management procedures) required to mobilize, organize and use these resources to confront the problems encountered and adapt to the reality created by hazard activity. Thirdly, [t]he planning and development strategies used to facilitate resilience must include mechanisms designed to integrate the resources available at each level to ensure the existence of a coherent societal capacity, and one capable of realizing the potential to capitalize on opportunities for change, growth and the enhancement of quality of life. Finally, strategies adapted must be designed to ensure the sustained availability of these resources and the competencies required to use them over time and against a background of hazard quiescence and changing community membership, needs, goals and functions.

9. See, for instance, McEntire ("Why Vulnerability Matters: Exploring the Merit of an Inclusive Disaster Reduction Concept," *Disaster Prevention and Management* 14, no. 2 (2005): 206–222) and Geis ("By Design: The Disaster-Resistant and Quality-of-Life Community," *Natural Hazards Review* 1, no. 3 (2000): 151–160).
10. For instance, see Smith (*Planning for Post-Disaster Recovery: A Review of the United States Disaster Assistance Framework* (Washington, DC: Island Press, 2012)) and Berke and Campanella ("Planning for Postdisaster Resiliency," *Annals of the American Academy of Political and Social Science* 604, Shelter from the Storm: Repairing the National Emergency Management System after Hurricane Katrina (2006): 192–207).
11. A large literature examines the psychological, social, and economic factors of community resilience and offers recommendations for how to foster resilience through infrastructure investment, community cohesion, and collective preparedness efforts (see Cohen et al., "The Conjoint Community Resiliency Assessment Measure as a Baseline for Profiling and Predicting Community Resilience for Emergencies," *Technological Forecasting and Social Change* 80, no. 9 (2013): 1732–1741; Kulig et al., "Community Resiliency: Emerging Theoretical Insights," *Journal of Community Psychology* 41, no. 6 (2013): 758–775; Norris et al., "Community Resilience as a Metaphor, Theory, Set of Capacities, and Strategy for Disaster Readiness," *American Journal of Community Psychology* 41, no. 1–2 (2008): 127–150; Adger et al., "Social-Ecological Resilience to Coastal Disasters," *Dealing with Disasters* 309, no. 5737 (2005): 1036–1039; Paton et al. (2000)).
12. Not surprisingly, entrepreneurship and adaptive capacity have been linked (see, for instance, Kirzner, *Competition and Entrepreneurship* (Chicago: University of Chicago Press, 1973); Etzioni, "Entrepreneurship, Adaptation and Legitimation: A Macro-Behavioral Perspective," *Journal of Economic Behavior and Organization* 8, no. 2 (1987): 175–189; Dernell et al., "The Economics

of Strategic Opportunity," *Strategic Management Journal* 24, no. 10 (2003): 977–990; Harrison and Leitch, "Entrepreneurial Learning: Researching the Interface between Learning and the Entrepreneurial Context," *Entrepreneurship Theory and Practice* 29, no. 4 (2005): 351–371).

13. A third policy recommendation might follow. Policymakers could also focus on providing the goods and services that entrepreneurs expect government to provide, and maybe providing some of the resources they need, but nothing else. Policymakers can encourage entrepreneurship after disasters by focusing on restoring needed utilities such as electricity and water, services such as policing and the courts, as well as clearing debris and reopening roads. After these basic services are restored, policymakers can focus on repairing important infrastructure, such as hospitals, schools, and highways. By promising to do relatively little, and keeping those promises, policymakers can provide a stable and predictable environment for entrepreneurs to thrive in.
14. Rigidly enforcing rules in this context could very well be a mistake.

Bibliography

ABC News, "Person of the Week: Doris Voitier." Last modified on August 25, 2006, http://abcnews.go.com/WNT/PersonOfWeek/story?id=2357707.

Aberbach, J. D., R. D. Putnam, and B. A. Rockman. *Bureaucrats and Politicians in Western Democracies* (Cambridge: Harvard University Press, 1981).

Abramson, L. "For One New Orleans School, an Uncertain Future." *National Public Radio.* Last modified May 17, 2007, http://www.npr.org/templates/story/story.php?storyId=10228892.

Adler, K. F. "Social Capital, Post-Disaster." In *Community Disaster Recover and Resilience*, ed. D. S. Miller and J. D. Rivera (Boca Raton: CRC Press, 2010): 125–173.

Adler, P. S. and S. W. Kwon. "Social Capital: Prospects for a New Concept." *Academy of Management Review* 27, no. 1 (2002): 17–40.

Adger, W. N. "Social and Ecological Resilience: Are They Related?" *Progress in Human Geography* 24, no. 3 (2000): 347–364.

Adger, W. N., T. P. Hughes, C. Folke, S. R. Carpenter, and J. Rockstrom. "Social-Ecological Resilience to Coastal Disasters." *Dealing with Disasters* 309, no. 5737 (2005): 1036–1039.

Advisory Commission on Intergovernmental Relations. *The Organization of Local Public Economies* (Washington, DC: Advisory Commission on Intergovernmental Relations, 1987).

Advisory Commission on Intergovernmental Regulations. *Metropolitan Organization: The St. Louis Case M-158* (Washington, DC: Advisory Commission on Intergovernmental Relations, 1988).

———. *Characteristics of Federal Grant-in-Aid Programs to State and Local Governments* (Washington, DC: Advisory Commission on Intergovernmental Relations, 1992).

Aldrich, D. P. "The Power of People: Social Capital's Role in Recovery from the 1995 Kobe Earthquake." *Natural Hazards* 56, no. 3 (2011a): 595–611.

———. "The Externalities of Strong Social Capital: Post-Tsunami Recovery in Southeast India." *Journal of Civil Society* 7, no. 1 (2011b): 81–99.

———. *Building Resilience: Social Capital in Post-Disaster Recovery* (Chicago: University of Chicago Press, 2012).

Aligica, P. and F. Sabetti. "Introduction: The Ostroms' Research Program for the Study of Institutions and Governance: Theoretical and Epistemic Foundations." In *Choice, Rules and Collective Action*, ed. P. Aligica and F. Sabetti (2014): 1–19.

Alvord, S. H., L. D. Brown, and C. W. Letts. "Social Entrepreneurship and Societal Transformation an Exploratory Study." *The Journal of Applied Behavioral Science* 40, no. 3 (2004): 260–282.

Arentz, J., F. Sautet, and V. Storr. "Prior-Knowledge and Opportunity Identification." *Small Business Economics* 41, no. 2 (2013): 461–478.

Austin, J., H. Stevenson, and J. Wei-Skillern. "Social and Commercial Entrepreneurship: Same, Different, or Both?" *Entrepreneurship Theory and Practice* 30, no. 1 (2006): 1–22.

Baldwin, W. L., J. Scott, and J. T. Scott. *Market Structure and Technological Change*. Fundamentals of Pure and Applied Economics Series 17 (London: Harwood Academic Publishers, 1987).

Bankston III, C. L. and M. Zhou. "De Facto Congregationalism and Socioeconomic Mobility in Laotian and Vietnamese Immigrant Communities: A Study of Religious Institutions and Economic Change." *Review of Religious Research* 41, no. 4 (2000): 453–470.

Baron, R. A. and G. D. Markman. "Cognitive Mechanisms: Potential Differences between Entrepreneurs and Non-entrepreneurs." *Frontiers of Entrepreneurship Research* (1999): 23–137.

———. "Beyond Social Capital: How Social Skills Can Enhance Entrepreneurs' Success." *Academy of Management Executive* 14, no. 1 (2000): 106–116.

Barrow, B. "Lack of Finances May Force Lower 9th Ward Health Clinic to Close Dec. 31." *The Times-Picayune*. Last modified December 14, 2010, http://www.nola.com/politics/index.ssf/2010/12/lack_of_finances_may_force_low.html.

Barsky, L., J. Trainor, and M. Torres. "Disaster Realities in the Aftermath of Hurricane Katrina: Revisiting the Looting Myth." *Natural Hazards Center Quick Response Report*, no. 184 (2006), http://udspace.udel.edu/bitstream/handle/19716/2367/Misc%20Report%2053.pdf?sequence=1.

Basurto, X. and E. Ostrom. "Beyond the Tragedy of the Commons." *Economia delle fonti di energia e dell'ambiente* 1 (2009): 35–60.

Bauer, J. "Market Power, Innovation, and Efficiency in Telecommunications: Schumpeter Reconsidered." *Journal of Economic Issues* 31, no. 2 (1997): 557–565.

Beito, D. T. *From Mutual Aid to the Welfare State: Fraternal Societies and Social Services, 1890–1967* (Chapel Hill, NC: University of North Carolina Press, 2002).

Bensoussan, B. "Mold, Rot & Truckloads of Goodwill." *Mishpacha 437*. Last modified December 3, 2012, http://www.mishpacha.com/Browse/Article/2718/Mold-Rot-and-Truckloads-of-Goodwill.

Berke, P. R. and T. J. Campanella. "Planning for Postdisaster Resiliency." *Annals of the American Academy of Political and Social Science* 604. Shelter from the Storm: Repairing the National Emergency Management System after Hurricane Katrina (2006): 192–207.

Besley, T. "Commodity Taxation and Imperfect Competition: A Note on the Effects of Entry." *Journal of Public Economics* 40, no. 3 (1989): 359–367.

Birch, E. and S. Wachter, eds. *Rebuilding Urban Places After Disaster: Lessons From Hurricane Katrina* (Philadelphia: University of Pennsylvania Press, 2006).

Bjørnskov, C. and N. Foss. "Economic Freedom and Entrepreneurial Activity: Some Cross-Country Evidence." *Public Choice* 134, no. 3/4 (2008): 307–328.

Blanchard-Boehm, R. D. "Understanding Public Response to Increased Risk from Natural Hazards: Application of the Hazards Risk Communication Framework." *International Journal of Mass Emergencies and Disasters* 16, no. 3 (1998): 247–278.

Boehm, S. "The Austrian Tradition: Schumpeter and Mises." In *Neoclassical Economic Theory 1870–1930*, ed. K. Hennings and W. Samuels (Boston: Kluwer Academic Publications, 1990): 201–249.

Boettke, P. J. and C. J. Coyne. "Entrepreneurship and Development: Cause or Consequence?" *Advances in Austrian Economics* 6 (2003): 67–88.

———. *Context Matters: Institutions and Entrepreneurship* (Hanover: Now Publishers Inc., 2009a).

———. "An Entrepreneurial Theory of Social and Cultural Change." In *Markets and Civil Society: The European experience in Comparative Perspective* (2009b): 77–103.

Boettke, P. J., C. J. Coyne, and P. T. Leeson. "Institutional Stickiness and the New Development Economics." *American Journal of Economics and Sociology* 67, no. 2 (2008): 331–358.

Boettke, P. J. and D. L. Prychitko. "Is an Independent Nonprofit Sector Prone to Failure?" *The Philanthropic Enterprise* 1, no. 1 (2004): 1–63.

Bogason, P. "The Fragmentation of Local Government in Scandinavia." *European Journal of Political Research* 30, no.1 (1996): 65–86.

Bolin, R. and L. Stanford. "The Northridge Earthquake: Community-based Approaches to Unmet Recovery Needs." *Disasters* 22, no. 1 (1998): 21–38.

Bornstein, D. *The Price of a Dream: The Story of the Grameen Bank and the Idea that is Helping the Poor to Change their Lives* (New York: Simon & Schuster, 1996).

Boudreaux, D. "Schumpeter and Kirzner on Competition and Equilibrium." In *The Market Process: Essays in Contemporary Austrian Economics*, ed. P. Boettke and D.L. Prychitko (Aldershot: Edward Elgar, 1994): 52–62.

Bourdieu, P. "The Social Space and the Genesis of Groups." *Theory and Society* 14, no. 6 (1985): 723–744.

Branzei, O. and S. Abdelnour. "Another Day, Another Dollar: Enterprise Resilience Under Terrorism in Developing Countries." *Journal of International Business Studies* 4, no. 5 (2010): 804–825.

Breschi, S., F. Malerba, and L. Orsenigo. "Technological Regimes and Schumpeterian Patterns of Innovation." *The Economic Journal* 110, no. 463 (2000): 388–410.

Brewer, M. "Planning Disaster: Price Gouging Statutes and the Shortages They Create." *Brooklyn Law Review* 72, no. 3 (2006–2007): 1101–1138.

Bronston, B. "Child-Care Industry Called 'Missing Piece' to Recovery; About 81% of Sites Suffered from Katrina." *The Times-Picayune*, April 28, 2006.

Brouwer, M. "Entrepreneurship and Uncertainty: Innovation and Competition Among the Many." *Small Business Economics* 15, no. 2 (2000): 149–160.

Brown, D. D. and J. C. Kulig. "The Concepts of Resiliency: Theoretical Lessons from Community Research." *Health and Canadian Society* 4, no. 1 (1996–1997): 29–50.

Buchanan, J. M. and G. Tullock, *The Calculus of Consent* (Indianapolis: Liberty Fund, 1999).

Bullough, A., M. Renko, and T. Myatt. "Danger Zone Entrepreneurs: The Importance of Resilience and Self-Efficiency for Entrepreneurial Intentions." *Entrepreneurship Theory and Practice* 38, no. 3 (2013): 473–499.

Burby, R. J. "Hurricane Katrina and the Paradoxes of Government Disaster Policy: Bringing about Wise Governmental Decisions for Hazardous Areas." *The Annals of the American Academy of Political and Social Science* 60, no. 1 (2006): 171–191.

Burt, R. *Structural Holes: The Social Structure of Competition* (Cambridge: Harvard University Press, 1992).

———. "The Gender of Social Capital." *Rationality and Society* 10, no. 1 (1998): 5–46.

Burton, C. G. "A Validation of Metrics for Community Resilience to Natural Hazards and Disasters Using the Recovery from Hurricane Katrina as a Case Study." *Annals of the Association of American Geographers* 105, no. 1 (2014): 67–86.

Cantrell, L. "Don't Sacrifice Our Homes to Flooding." *New Orleans Times Picayune*. Last modified March 30, 2006, https://repository.library.brown.edu/studio/item/bdr:65484/.

Carden, A., C. Courtemanche, and J. Meiners. "Does Wal-Mart Reduce Social Capital?" *Public Choice* 138, no. 1–2 (2009): 109–136.

Chamlee-Wright, E. "Local Knowledge and the Philanthropic Process: Comment on Boettke and Prychitko." *Conversations in Philanthropy* 1, no. 1 (2004): 45–51.

———. "The Long Road Back: Signal Noise in the Post-Katrina Context." *The Independent Review* 12, no. 2 (2007): 235–259.

———. "The Structure of Social Capital: An Austrian Perspective on Its Nature and Development." *Review of Political Economy* 20, no. 1 (2008): 41–58.

———. *The Cultural and Political Economy of Recovery: Social Learning in a Post-Disaster Environment* (London: Routledge, 2010).

———. "Pastor Response in Post-Katrina New Orleans: Navigating the Cultural Economic Landscape." In *Culture and Economic Action*, ed. L. E. Grube and V. H. Storr (Northampton: Edward Elgar 2015: 269–294.)

Chamlee-Wright, E. and J. A. Myers. "Discovery and Social Learning in Non-priced Environments: An Austrian View of Social Network Theory." *The Review of Austrian Economics* 21, no. 2–3 (2008): 151–166.

Chamlee-Wright, E. and V. H. Storr. "The Entrepreneur's Role in Post Disaster Community Recovery: Implications for Post Disaster Recovery Policy." *Mercatus Center Policy Series*, Primer No. 6 (Arlington, VA: Mercatus Center at George Mason University, September 2008).

———. "Club Goods and Post-Disaster Community Return." *Rationality and Society* 21, no. 4 (2009a): 429–458.

———. "There's No Place Like New Orleans: Sense of Place and Community Recovery in the Ninth Ward after Hurricane Katrina." *Journal of Urban Affairs* 31, no. 5 (2009b): 615–634.

———. "Filling the Civil Society Vacuum: Post Disaster Policy and Community Response." *Mercatus Center Policy*, Policy Comment No. 2 (Arlington, VA: Mercatus Center at George Mason University, February 2009c).

———. "The Role of Social Entrepreneurship in Post-Katrina Community Recovery." *International Journal of Innovation and Regional Development* 2, no. 1 (2010a): 149–164.

———. "Expectations of Government's Response to Disaster." *Public Choice* 144, no. 1–2 (2010b): 253–274.

———. "Social Capital, Lobbying and Community-based Interest Groups." *Public Choice* 149, no. 1–2 (2011a): 167–185.

———. "Social Capital as Collective Narratives and Post Disaster Community Recovery." *The Sociological Review* 59, no. 2 (2011b): 266–282.

———. "Community Resilience in New Orleans East: Deploying the Cultural Toolkit Within a Vietnamese-American Community." In *Community Disaster Recovery and Resiliency: Exploring Global Opportunities and Challenges*, ed. J. Rivera and D. Miller (Boca Raton: Taylor and Francis Group, 2011c): 101–124.

———. "Commercial Relationships and Spaces after Disaster." *Society* 51, no. 6 (2014): 656–664.

Choi, Y. B. "The Entrepreneur: Schumpeter vs. Kirzner." In *Advances in Austrian Economics* 2(A), ed. P. Boettke and M. Rizzo (Greenwich: JAI Press, 1995): 55–78.

Chong, A., J. Guillen, and V. Rios. *The Cultural and Political Economy of Recovery: Social Learning in a Post Disaster Environment* (London: Routledge, 2010).

Cigler, B. A. "Post-Katrina Hazard Mitigation on the Gulf Coast." *Public Organization Review* 9, no. 4 (2009): 325–341.

Cohen, O., D. Leykin, M. Lahad, A. Goldberg, and L. Aharonson-Daniel. "The Conjoint Community Resiliency Assessment Measure as a Baseline for Profiling and Predicting Community Resilience for Emergencies." *Technological Forecasting and Social Change* 80, no. 9 (2013): 1732–1741.

Coleman, J. S. "Social Capital in the Creation of Human Capital." *American Journal of Sociology* 94 (1988): S95-S120.

Coppola, D. *Introduction to International Disaster Management* (Oxford: Elsevier, 2015).

Corbett, A. C. "Learning Asymmetries and the Discovery of Entrepreneurial Opportunities." *Journal of Business Venturing* 22, no. 1 (2007): 97–118.

Coyne, C. J. "The Importance of Expectations in Economic Development." *The Wealth and Well-Being of Nations* 1 (2009): 63–82.

———. *Doing Bad by Doing Good* (California: Stanford University Press, 2013).

Coyne, C. J. and J. Lemke. "Polycentricity in Disaster Relief." *Studies in Emergent Order* 4 (2011): 40–57.

———. "Lessons from the Cultural and Political Economy of Recovery." *American Journal of Economics and Sociology* 71, no. 1 (2012): 215–228.

Crabill, A. and Y. Rademacher. "Whole Community: Local, State, and Federal Relationships." In *Critical Issues in Disaster Science and Management: A Dialogue between Researchers and Practitioners*, ed. J. Trainor and T. Subbio (United States Federal Emergency Management Agency, 2012): 9–53.

Cutter, S. L., C. G. Burton, and C. T. Emrich. "Disaster Resilience Indicators for Benchmarking Baseline Conditions." *Journal of Homeland Security and Emergency Management* 7, no. 1 (2010): 1–22.

Dacin, P. A., M. T. Dacin, and M. Matear. "Social Entrepreneurship: Why We Don't Need a New Theory and How We Move Forward from Here." *Academy of Management Perspectives* 24, no. 3 (2010): 37–57.

Dahl, Robert A. *Who Governs? Democracy and Power in an American City* (New Haven: Yale University Press, 1961).

Darling-Hammond, L. "Race, Inequality and Educational Accountability: The Irony of 'No Child Left Behind.'" *Race Ethnicity and Education* 10, no. 3 (2007): 245–260.

Dees, J. G. "The Meaning of Social Entrepreneurship" (1998), http://csi.gsb.stanford.edu/sites/csi.gsb.stanford.edu/files/TheMeaningofsocialEntrepreneurship.pdf.

Dees, J. G. and P. Economy. "Social Entrepreneurship." In *Enterprising Nonprofits: A Toolkit for Social Entrepreneurs*, ed. J. G. Dees, J. Emerson, and P. Economy (New York: John Wiley and Sons, 2001): 1–18.

Delta Sky, "Life in the New Normal." Last modified January 2008.

Denrell, J., C. Fang, and S. G. Winter. "The Economics of Strategic Opportunity." *Strategic Management Journal* 24, no. 10 (2003): 977–990.

Denzau, A. and D. North. "Shared Mental Models: Ideologies and Institutions." *Kyklos* 47, no. 1 (1994): 3–31.

d'Hombres, B., L. Rocco, M. Suhrcke, and M. McKee. "Does Social Capital Determine Health? Evidence from Eight Transition Countries." *Health Economics* 19, no. 1 (2010): 56–74.

Dilulio, J. J. "Help Wanted: Economists, Crime and Public Policy." *The Journal of Economic Perspectives* 10, no. 1 (1996): 93–24.

DiLorenzo, T. J. "Competition and Political Entrepreneurship: Austrian Insights into Public-Choice Theory." *The Review of Austrian Economics* 2, no. 1 (1988): 59–71.

DiMaggio, P. "The New Institutionalisms: Avenues of Collaboration." *Journal of Institutional and Theoretical Economics* 154, no. 4 (1998): 696–705.

Dimov, D. "From Opportunity Insight to Opportunity Intention: The Importance of Person–Situation Learning Match." *Entrepreneurship Theory and Practice* 31, no. 4 (2007): 561–583.

Dowla, A. "In Credit We Trust: Building Social Capital by Grameen Bank in Bangladesh." *The Journal of Socio-Economics* 35, no. 1 (2006): 102–122.

Drucker, P. F. "Entrepreneurial Strategies." *California Management Review* 27, no. 2 (1985): 9–25.

Dunne, M. "Failure to Halt Landfill Doesn't Stop Activists." *The Advocate [Baton Rouge]*. Last modified April 28, 2006.

Dynes, R. R. "Social Capital: Dealing with Community Emergencies." *Homeland Security Affairs* 2, no. 2 (2006): 1–26.

Eisenstadt, S. N. "Cultural Orientations, Institutional Entrepreneurs, and Social Change: Comparative Analysis of Traditional Civilizations." *American Journal of Sociology* 85, no. 4 (1980): 840–869.

Elkington, J. and P. Hartigan. *The Power of Unreasonable People: How Social Entrepreneurs Create Markets That Change the World* (Boston: Harvard Business Review Press, 2008).

Emerson, J. and F. Twersky, eds. *New Social Entrepreneurs: The Success, Challenge, and Lessons of Non-profit Enterprise Creation* (San Francisco: Roberts Foundation, 1996).

Endres, A. M. and C. R. Woods. "Modern Theories of Entrepreneurial Behavior: A Comparison and Appraisal." *Small Business Economics* 26, no. 2 (2006): 189–202.

Enterprise Partners, "Hurricane Sandy: Housing Needs One Year Later." October 2013, https://s3.amazonaws.com/KSPProd/ERC_Upload/0083708.pdf.

Erikson, K. T. *Everything in Its Path: Destruction of Community in the Buffalo Creek Flood* (New York: Simon and Schuster, 1976).

Etheridge, F., I. McNulty, A. Norris, and D. Winkler-Schmit. "Gambit's 40 Under 40." *Gambit*. Last modified November 07, 2006, http://www.bestofneworleans.com/gambit/40-under-40/Content?oid=1246675.

Etzioni, A. "Entrepreneurship, Adaptation and Legitimation: A Macro-Behavioral Perspective." *Journal of Economic Behavior and Organization* 8, no. 2 (1987): 175–189.

Fakhruddin, S. H. M. and Y. Chivakidakarn. "A Case Study for Early Warning and Disaster Management in Thailand." *International Journal of Disaster Risk Reduction* 9 (2014): 159–180.

Foley, M. W. and B. Edwards. "Is It Time to Disinvest in Social Capital?" *Journal of Public Policy* 19, no. 2 (1999): 141–173.

Folson, B. W. *Entrepreneurs vs. the State: A New Look at the Rise of Big Business in America* (Reston: Young America's Foundation, 1987).

Foss, N. J., ed. *Resources, Firms, and Strategies: A Reader in the Resource-Based Perspective* (New York: Oxford University Press, 1997).

Fothergill, A. and Peek, L. "Poverty and Disasters in the United States: A Review of Recent Sociological Findings." *Natural Hazards* 32, no. 1 (2004): 89–110.

Fox News, "FOX Facts: Hurricane Katrina Damage." Last modified Aug. 29, 2006, http://www.foxnews.com/story/2006/08/29/fox-facts-hurricane-katrina-damage/.

Fukuyama, F. "Social Capital, Civil Society and Development." *Third World Quarterly* 22, no. 1 (2001): 7–20.

Fussell, E. "Help from Family, Friends, and Strangers During Hurricane Katrina: Finding the Limits of Social Networks." In *Displaced: Life in the Katrina Diaspora*, ed. L. Weber and L. Peek (Austin: University of Texas Press, 2012): 150–166.

Gaglio, C. M. "The Role of Mental Simulations and Counterfactual Thinking in the Opportunity Identification Process." *Entrepreneurship Theory and Practice* 28, no. 6 (2004): 533–552.

Gaglio, C. M. and J. A. Katz. "The Psychological Basis of Opportunity Identification: Entrepreneurial Alertness." *Small Business Economics* 16, no. 2 (2001): 95–111.

Galbraith, C. S. and C. H. Stiles. "Disasters and Entrepreneurship: A Short Review." *International Research in the Business Disciplines* 5 (2006): 147–166.

Garrett, T. A. and R. S. Sobel. "The Political Economy of FEMA Disaster Payments." *Economic Inquiry* 41, no. 3 (2003): 496–509.

Gartner, W. B. "What are We Talking about When We Talk about Entrepreneurship?" *Journal of Business Venturing* 5, no. 1 (1990): 15–28.

Geaghan, K. A. "Forced to Move: An Analysis of Hurricane Katrina Movers." *Social, Economic, and Housing Statistics Division.* US Census Bureau Working Paper, 2011.

Geis, D. "By Design: The Disaster-Resistant and Quality-of-Life Community." *Natural Hazards Review* 1, no. 3 (2000): 151–160.

Gerteis, J. "The Possession of Civic Virtue: Movement Narratives of Race and Class in the Knights of Labor." *American Journal of Sociology* 108, no. 3 (2002): 580–615.

Giersch, H. "The Age of Schumpeter." *The American Economic Review* 74, no. 2 (1984): 103–109.

Goetz, A. M. and R. S. Gupta. "Who Takes the Credit? Gender, Power, and Control over Loan Use in Rural Credit Programs in Bangladesh." *World Development* 24, no. 1 (1996): 45–63.

Gosselin, P. G. "On Their Own in Battered New Orleans." *Los Angeles Times.* Last modified December 04, 2006, http://articles.latimes.com/2005/dec/04/nation/na-orleansrisk4.

Granovetter, M. "The Strength of Weak Ties." *American Journal of Sociology* 78, no. 6 (1973): 1360–1380.

———. "The Strength of Weak Ties: A Network Theory Revisited." *Sociological Theory* 1, no. 1 (1983): 201–233.

———. "Economic Action and Social Structure: The Problem of Embeddedness." *American Journal of Sociology* 91, no. 3 (1985): 481–510.

———. *Getting a Job: A Study of Contacts and Careers* (Chicago: University of Chicago Press, 1995).

Grube, L. and V. H. Storr. "The Capacity for Self-Governance and Post Disaster Resiliency." *The Review of Austrian Economics* 27, no. 3 (2014): 301–324.

Guisbond, L. and M. Neill. "Failing Our Children: No Child Left Behind Undermines Quality and Equity in Education." *The Clearing House: A Journal of Educational Strategies, Issues, and Ideas* 78, no. 1 (2004): 12–16.

Gyan, J. "Few N.O. Day-Care Centers Running." *The Advocate*, November 3, 2007.

Hall, P. A. "Policy Paradigms, Social Learning, and the State: The Case of Economic Policymaking in Britain." *Comparative Politics* 25, no. 3 (1993): 275–296.

Hanushek, Eric A. "The Failure of Input-based Schooling Policies." *The Economic Journal* 113, no. 485 (2003): F64-F98.

Harrison, R. T. and C. M. Leitch. "Entrepreneurial Learning: Researching the Interface between Learning and the Entrepreneurial Context." *Entrepreneurship Theory and Practice* 29, no. 4 (2005): 351–371.

Hassan, K. M. "The Microfinance Revolution and the Grameen Bank Experience in Bangladesh." *Financial Markets, Institutions & Instruments* 11, no. 3 (2002): 205–265.

Hay, C. "Narrating Crisis: The Discursive Construction of the Winter of Discontent." *Sociology* 30, no. 2 (1996): 253–277.

Hayek, F. A. "The Use of Knowledge in Society." *The American Economic Review* 35, no. 4 (1945): 519–530.

———., ed. *Individualism and Economic Order* (Chicago: University of Chicago Press, 1948).

———. "Degrees of Explanation." *The British Journal for the Philosophy of Science* 6, no. 23 (1955): 209–225.

Hébert, R. F. and A. N Link. *The Entrepreneur: Mainstream Views and Radical Critiques* (New York: Praeger, 1982).

———. "In Search of the Meaning of Entrepreneurship." *Small Business Economics* 1, no. 1 (1989): 39–49.

Higgs, R. "Regime Uncertainty." *Independent Review* 1, no. 4 (1997): 561–590.

Hills, G. E. and R. C. Shrader. "Successful Entrepreneurs' Insights into Opportunity Recognition." *Frontiers of Entrepreneurship Research* 18 (1998): 30–41.

Holcombe, R. "Political Entrepreneurship and the Democratic Allocation of Economic Resources." *The Review of Austrian Economics* 15, no. 2–3 (2002): 143–159.

———. *Entrepreneurship and Economic Progress* (New York: Routledge, 2007).

Holling, C. S. "Resilience and Stability of Ecological Systems." *Annual Review of Ecology and Systematics* 4 (1973): 1–23.

Horwitz, S. "Wal-Mart to the Rescue: Private Enterprise's Response to Hurricane Katrina." *The Independent Review* 13, no. 4 (2009a): 511–528.

———. "Best Responders: Post-Katrina Innovation and Improvisation by Wal-Mart and the U.S. Coast Guard." *Innovations* 4, no. 2 (2009b): 93–99.

Hurlbert, J. S., J. J. Beggs, and V. A. Haines. "Core Networks and Tie Activation: What Kinds of Routine Networks Allocate Resources in Nonroutine Situations?" *American Sociological Review* 65, no. 4 (2000): 598–618.

———. "Social Networks and Social Capital in Extreme Environments." In *Social Capital: Theory and Research*, ed. N. Lin, K. S. Cook, and R. S. Burt (Cambridge: Cambridge University Press, 2001): 209–231.

Hwang, H. and W. W Powell. "Institutions and Entrepreneurship." In *Handbook of Entrepreneurship Research*, ed. S. A. Alvarez, R. Agarwal, and O. Sorenson (New York: Springer, 2005): 201–232.

Ikeda, S. and P. Gordon. "Power to the Neighborhoods: The Devolution of Authority in Post-Katrina New Orleans." *Mercatus Policy Series*, Policy Comment No. 12 (Arlington, VA: Mercatus Center at George Mason University, August 2007).

Irazábal, C. and J. Neville. "Neighborhoods in the Lead: Grassroots Planning for Social Transformation in Post-Katrina New Orleans." *Planning, Practice & Research* 22, no. 2 (2007): 131–153.

Jack, B. K. and M. P. Recalde. *Local Leadership and the Voluntary Provision of Public Goods: Field Evidence from Bolivia*, Working Paper, 2013.

Jacobs, J. *The Death and Life of Great American Cities* (New York: Random House Inc., 1961).

Jain, P. S. "Managing Credit for the Rural Poor: Lessons from the Grameen Bank." *World Development* 24, no. 1 (1996): 79–89.

Jaycox, L., J. Cohen, J. Mannarino, D. Walker, A. Langley, K. Gegenheimer, M. Scott, and M. Schonlau. "Children's Mental Health Following Hurricane Katrina: A Field Trial of Trauma-Focused Psychotherapies." *Journal of Traumatic Stress* 23, no. 2 (2010): 223–231.

Jones C. *Introduction to Economic Growth* (New York: W.W. Norton & Company, 1998).

Joseph, S. and P. Linley, eds. *Trauma, Recovery, and Growth: Positive Psychological Perspectives on Posttraumatic Stress* (Hoboken: Wiley, 2008).

Kabir Hassan, M. "The Microfinance Revolution and the Grameen Bank Experience in Bangladesh." *Financial Markets, Institutions & Instruments* 11, no. 3 (2002): 205–265.

Kahn, Matthew E. "The Death Toll from Natural Disasters: The Role of Income, Geography, and Institutions." *Review of Economics and Statistics* 87, no. 2 (2005): 271–284.

Kaish, S. and B. Gilad. "Characteristics of Opportunities Search of Entrepreneurs Versus Executives: Sources, Interests, General Alertness." *Journal of Business Venturing* 6, no. 1 (1991): 45–61.

Kaplan, L. and C. Kaplan. *Between Ocean and City: The Transformation of Rockaway, New York* (New York: Columbia University Press, 2003).

Keh, H. T., M. D. Foo, and B. C. Lim. "Opportunity Evaluation under Risky Conditions: The Cognitive Processes of Entrepreneurs." *Entrepreneurship Theory and Practice* 27, no. 2 (2002): 125–148.

Kessler, R., S. Galea, M. Gruber, N. Sampson, R. Ursano, and S. Wessely. "Trends in Mental Illness and Suicidality after Hurricane Katrina." *Molecular Psychiatry* 13 (2008): 374–384.

King, D. "Organizations in Disaster." *Natural Hazards* 40, no. 3 (2007): 657–665.

Kirzner, I. M. *Competition and Entrepreneurship* (Chicago: University of Chicago Press, 1973).

———. *Perception, Opportunity, and Profit: Studies in the Theory of Entrepreneurship* (Chicago: University of Chicago Press, 1979).

———. "Entrepreneurs and the Entrepreneurial Function: A Commentary." In *Entrepreneurship: Where Did It Come From, and Where Is It Going*, ed. J. Ronen (Lexington: D. C. Heath, 1983): 281–290.

———. *Discovery and the Capitalist Process* (Chicago: University of Chicago Press, 1985).

———. "The Austrian School of Economics." In *New Palgrave Dictionary of Economics*, ed. S. N. Durlauf and L. E. Blume (New York: Macmillan, 1986).

———. "Entrepreneurship." In *The Elgar Companion to Austrian Economics*, ed. P. Boettke (Northampton: Edward Elgar Publishing Inc., 1994): 103–111.

———. "Entrepreneurial Discovery and the Competitive Market Process: An Austrian Approach." *Journal of Economic Literature* 35, no. 1 (1997): 60–85.

———. "Creativity and/or Alertness: A Reconsideration of the Schumpeterian Entrepreneur." *The Review of Austrian Economics* 11, no. 1 (1999): 5–17.

———. *The Driving Force of the Market: Essays in Austrian Economics* (London: Routledge, 2000).

———. "The Alert and Creative Entrepreneur: A Clarification." *Small Business Economics* 32, no. 2 (2009): 145–152.

Kirzner, I. M. and F. Sautet. "The Nature and Role of Entrepreneurship in Markets: Implications for Policy." *Mercatus Policy Series Policy Primer*, No. 4 (Arlington, VA: Mercatus Center at George Mason University, June 2006).

Klein, D. B. *Knowledge and Coordination: A Liberal Interpretation* (New York: Oxford University Press, 2012).

Klein, P. G. and P. L. Bylund. "The Place of Austrian Economics in Contemporary Entrepreneurship Research." *The Review of Austrian Economics* 27, no. 3 (2014): 259–279.

Knabb, R. D., J. R. Rhome, and D. P. Brown, "Tropical Cyclone Report: Hurricane Katrina, 23–20 August 2005." *National Hurricane Center*. Last modified August 10, 2006, http://www.nhc.noaa.gov/data/tcr/AL122005_Katrina.pdf

Knack, S. and P. Keefer. "Does Social Capital Have an Economic Payoff? A Cross-Country Investigation." *The Quarterly Journal of Economics* 112, no. 4 (1997): 1251–1288.

Knight, F. H. *Risk, Uncertainty and Profit* (New York: Hart, Schaffner, and Marx, 1921).

———. "Profit and Entrepreneurial Functions." *The Journal of Economic History* 2 (1942): 126–132.

Krane, D. A. "The State of American Federalism, 2001–2002: Resilience in Response to Crisis." *Publius: The Journal of Federalism* 32, no. 4 (2002): 1–28.

Kraft, K. "Market Structure, Firm Characteristics and Innovative Activity." *The Journal of Industrial Economics* 37, no. 3 (1989): 329–336.

Kulig, J. C. "Community Resiliency: The Potential for Community Health Nursing Theory Development." *Public Health Nursing* 17, no. 5 (2000): 374–385.

Kulig, J. C., D. S. Edge, I. Townshend, N. Lightfoot, and W. Reimer. "Community Resiliency: Emerging Theoretical Insights." *Journal of Community Psychology* 41, no. 6 (2013): 758–775.

Lachmann, L. M. *Capital, Expectations and the Market Process* (Kansas City: Sheed and McMeel, 1977).

———. "An Austrian Stocktaking: Unsettled Questions and Tentative Answers." In *New Directions in Austrian Economics*, ed. Louis M. Spadaro (Kansas City: Sheed Andrews and McMee, 1978): 1–18.

———. *The Market as an Economic Process* (Oxford: Basil Blackwell, 1986).

———. "Austrian Economics as a Hermeneutic Approach." In *Economics and Hermeneutics*, ed. Don Lavoie (London: Routledge, 1991): 134–146.

Landphair, J. "The Forgotten People of New Orleans: Community, Vulnerability, and the Lower Ninth Ward." *The Journal of American History* 94, no. 3 (2007): 837–845.

Larance, L. Y. "Fostering Social Capital Through NGO Design Grameen Bank Membership in Bangladesh." *International Social Work* 44, no. 1 (2001): 7–18.

Leca, B., J. Battilana, and E. Boxenbaum. *Agency and Institutions: A Review of Institutional Entrepreneurship* (Cambridge: Harvard Business School Press, 2008).

Leeson, P. T. and P. J. Boettke. "The Austrian School of Economics: 1950–2000." In *A Companion to the History of Economic Thought*, ed. Warren Samuels, Jeff Biddle, and John Davis (Oxford: Blackwell Publishing Ltd., 2003): 445–453.

———. "Two-Tiered Entrepreneurship and Economic Development." *International Review of Law and Economics* 29, no. 3 (2009): 252–259.

Leeson, P. T. and R. S. Sobel. "Weathering Corruption." *Journal of Law and Economics* 51, no. 4 (2008): 667–681.

Leibtag, C. "Young Israel COO Testifies at NYC Council Hearing about Hurricane Sandy Relief." *Matzav Newscenter*, Feb. 7, 2013, http://matzav.com/young-israel-coo-testifies-at-nyc-council-hearing-about-hurricane-sandy-relief.

Light, P. C. "Reshaping Social Entrepreneurship." *Stanford Social Innovation Review* (2006): 47–51.

Lightman, H. "Voices from the Storm." *Jewish Action,* March 4, 2013, https://www.ou.org/jewish_action/03/2013/hurricane-sandy-a-spiritual-response/.

Maclaurin, W. R. "Technological Progress in Some American Industries." *The American Economic Review* 44, no. 2 (1954): 178–189.

Mair, J., J. Robinson, and K. Hockerts, eds. *Social Entrepreneurship* (New York: Palgrave Macmillan, 2006).

Markman, G., R. Baron, and D. Balkin. "Are Perseverance and Self-Efficacy Costless? Assessing Entrepreneurs' Eegretful Thinking." *Journal of Organizational Behavior* 26, no. 1 (2005): 1–19.

Martin, A. "Emergent Politics and the Power of Ideas." *Studies in Emergent Order* 3 (2010): 212–245.

Martin, N. P. and V. H. Storr. "I'se a Man: Political Awakening and the 1942 Riot in the Bahamas." *The Journal of Caribbean History* 41, no. 1/2 (2007): 72–91.

———. "Demystifying Bay Street: Black Tuesday and the Radicalization of Bahamian Politics in the 1960s." *The Journal of Caribbean History* 43, no. 1 (2009): 37–50.

Martin, R. L. and S. Osberg. "Social Entrepreneurship: The Case for Definition." *Stanford Social Innovation Review* (2007): 29–39.

McCloskey, D. N. "A Kirznerian Economic History of the Modern World." *Annual Proceedings of the Wealth and Well-Being of Nations* 3 (2010–2011): 45–64.

McEnery, T. "Oakwood Beach: Sell Out, Tear Down and Leave." *Crain's New York Business.* Last modified October 20, 2014, http://www.crainsnewyork.com/article/20141020/REAL_ESTATE/310199992/oakwood-beach-sell-out-tear-down-and-leave.

McEntire, D. A. "Why Vulnerability Matters: Exploring the Merit of an Inclusive Disaster Reduction Concept." *Disaster Prevention and Management* 14, no. 2 (2005): 206–222.

McGinnis, M. D., ed. *Polycentric Governance and Development: Readings from the Workshop in Political Theory and Policy Analysis* (Ann Arbor: University of Michigan Press, 1999a).

———., ed. *Polycentricity and Local Public Economies: Readings from the Workshop in Political Theory and Policy Analysis* (Ann Arbor: University of Michigan Press, 1999b).

———. "Introduction." In *Polycentricity and Local Public Economies: Readings from the Workshop in Political Theory and Policy Analysis*, ed. M. D. McGinnis (Ann Arbor: University of Michigan Press, 1999c): 1–27.

———., ed. *Polycentric Games and Institutions: Readings from the Workshop in Political Theory and Policy Analysis* (Ann Arbor: University of Michigan Press, 2000).

McGinnis, M. D. and J. M. Walker. "Foundations of the Ostrom Workshop: Institutional Analysis, Polycentricity, and Self-Governance of the Commons." *Public Choice* 143 (2010): 293–301.

Meadowcroft, J. and M. Pennington. "Bonding and Bridging: Social Capital and the Communitarian Critique of Liberal Markets." *The Review of Austrian Economics* 21, no. 2–3 (2008): 119–133.

Minniti, M. and R. Koppl. "Market Processes and Entrepreneurial Studies." In *Handbook of Entrepreneurship Research*, ed. Z. Acs and D. Audretsch (London: Kluwer Press International, 2003): 81–102.

Mises, L. V. *Bureacracy* (Indianapolis: Liberty Fund [1944] 2007).

———. *Human Action: A Treatise on Economics* (New Haven: Yale University Press, [1949] 1963).

———. *Economic Calculation in the Socialist Commonwealth* (Ludwig von Mises Institute, [1920] 1990).

Morduch, J. "The Role of Subsidies in Microfinance: Evidence from the Grameen Bank." *Journal of Development Economics* 60, 1 (1999): 229–248.

Moreno, R. *Firm Competitive Strategies and the Likelihood of Survival. The Spanish Case* (Jena: Max Planck Institute of Economics, Entrepreneurship, Growth and Public Policy Group, 2007).

Mort, G., J. Weerawardena, and K. Carnegie. "Social Entrepreneurship: Towards Conceptualization." *International Journal of Non-profit and Voluntary Sector Marketing* 8 (2003): 76–88.

Mullins, J. W. and Forlani, D. "Missing the Boat or Sinking the Boat: A Study of New Venture Decision Making." *Journal of Business Venturing* 20, no. 1 (2005): 47–69.

Murphy, B. L. "Locating Social Capital in Resilient Community-Level Emergency Management." *Natural Hazards* 41, no. 2 (2007): 297–315.

Nagle, D. "Everyday Heroes: Kasim, Owner of the Wash Zone, New Orleans." *American Coin-Op*. Last modified July 30, 2008, https://americancoinop.com/articles/everyday-heroes-kasim-owner-wash-zone-new-orleans.

Nakagawa, Y. and R. Shaw. "Social Capital: A Missing Link to Disaster Recovery." *International Journal of Mass Emergencies and Disasters* 22, no. 1 (2004): 5–34.

Naughton, M. J. and J. R. Cornwall. "The Virtue of Courage in Entrepreneurship: Engaging the Catholic Social Tradition and the Life-Cycle of the Business." *Business Ethics Quarterly* 16, no. 1 (2006): 69–93.

Nelson, O. "The Chicago Relief and Aid Society 1850–1874." *Journal of the Illinois State Historical Society (1908–1984)* 59, no. 1 (1966): 48–66.

Nicholson, P. "Hurricane Katrina: Why Did the Levees Fail?" United States Senate, One Hundred Ninth Congress, First Session, November 2 2005, http://www.gpo.gov/fdsys/pkg/CHRG-109shrg24446/html/CHRG-109shrg24446.htm.

Nolan, B. "6 Years Later, Hurricane Katrina's Scars Linger Alongside Robust Recovery." *The Times-Picayune*. Last modified Aug. 28, 2011, http://www.nola.com/katrina/index.ssf/2011/08/6_years_later_hurricane_katrin.html.

Norris, F. H., S. P. Stevens, B. Pfefferbaum, K. F. Wyche, and R. L. Pfefferbaum. "Community Resilience as a Metaphor, Theory, Set of Capacities, and Strategy for Disaster Readiness." *American Journal of Community Psychology* 41, no. 1–2 (2008): 127–150.

Norris, F. H., J. L. Perilla, J. K. Riad, K. Kaniasty, and E. A. Lavizzo. "Stability and Change in Stress, Resources, and Psychological Distress Following Natural Disaster: Findings from Hurricane Andrew." *Anxiety, Stress & Coping* 12, no. 4 (1999): 363–396.

North, D. *Structure and Change in Economic History* (New York: W. W. Norton, 1981).

NPR Religion, "Turning Toward Faith during Hurricane Aftermath." Last modified Aug. 29, 2011, http://www.npr.org/2011/08/29/140036221/turning-toward-faith-during-hurricane-aftermath.

Oakerson, R. J., S. Wynne, T. V. Truong, and S. T. Walker. "Privatization Structures: An Institutional Analysis of the Fertilizer Sub-sector Reform Program in Cameroon." *A report prepared for the Private Enterprise Development Support Project and I USAID/Cameroon* (Washington, DC: Ernst and Young, 1990).

O'Driscoll, G. and M. Rizzo. *The Economics of Time and Ignorance* (Oxford: Basil Blackwell, 1985).

Olowu, D. and J. Wunsch. *Local Governance in Africa: The Challenges of Democratic Decentralization* (Boulder: Lynne Rienner Publishers, 2004).

Olshansky, Robert B., L. A. Johnson, J. Horne, and B. Nee. "Longer View: Planning for the Rebuilding of New Orleans." *Journal of the American Planning Association* 74, no. 3 (2008): 273–287.

Ostrom, E. *Public Entrepreneurship: A Case Study in Ground Water Basin Management*, Doctoral dissertation (University of California, Los Angeles, 1965).

———. *Governing the Commons: The Evolution of Institutions For Collective Action* (Cambridge: Cambridge University Press, 1990).

———. "The Rudiments of a Theory of the Origins, Survival, and Performance of Common Property Institutions." In *Making the Commons Work: Theory, Practice, and Policy*, ed. D. W. Bromley, D. Feeny, M. A. McKean, P. Peters, J. L. Gilles, R. J. Oakerson, C. F. Runge, and J. T. Thomson (San Francisco: Institute for Contemporary Studies Press, 1992): 293–318.

———. "Constituting Social Capital and Collective Action." *Journal of Theoretical Politics* 6, no. 4 (1994): 527–562.

———. "Coping with Tragedies of the Commons." *Annual Review of Political Science* 2, no. 1 (1999): 493–535.

———. "Crowding Out Citizenship." *Scandinavian Political Studies* 23, no. 1 (2000): 3–16.

———. "A Diagnostic Approach for Going Beyond Panaceas." *Proceedings of the National Academy of Sciences* 104, no. 39 (2007): 15181–15187.

———. "Collective Action and the Evolution of Social Norms." *Journal of Natural Resources Policy Research* 6, no. 4 (2014): 235–252.

Ostrom, E. and M. Cox. "Moving Beyond Panaceas: A Multi-tiered Diagnostic Approach for Social-Ecological Analysis." *Environmental Conservation* 37, no. 4 (2010): 451–463.

Ostrom, E., M. Janssen, and J. Anderies. "Going Beyond Panaceas." *Proceedings of the National Academy of Sciences* 104, no. 39 (2007): 15176–15178.

Ostrom, V. "Polycentricity (Part 1)." In *Polycentricity and Local Public Economies: Readings from the Workshop in Political Theory and Policy Analysis*, ed. M. D. McGinnis (Ann Arbor: University of Michigan Press, [1972] 1999): 52–74.

———. "Polycentricity: The Structural Basis of Self-Governing Systems." In *Choice, Rules and Collective Action*, ed. P. Aligica and F. Sabetti ([1991] 2014): 45–60.

Ostrom, V. and E. Ostrom. "A Theory for Institutional Analysis of Common Pool Problems." In *Managing the Commons*, ed. G. Hardin and J. Baden (San Francisco: W.H. Freeman, 1977): 157–172.

———. "Public Goods and Public Choices." In *Polycentricity and Local Public Economies. Readings from the Workshop in Political Theory and Policy Analysis*, ed. M. McGinnis (Ann Arbor: University of Michigan Press, [1977] 1999): 75–103.

———. "Public Choice: A Different Approach to the Study of Public Administration." In *Choice, Rules and Collective Action*, ed. P. Aligica and F. Sabetti ([1971] 2014): 23–44.

Ostrom, V., C. M. Tiebout, and R. Warren. "The Organization of Government in Metropolitan Areas: A Theoretical Inquiry." American *Political Science Review* 55, no. 4 (1961): 831–842.

Ostrom, V., R. L. Bish, and E. Ostrom. *Local Government in the United States* (New York: ICS Press, 1988).

Paton, D. "Disaster Resilience: Building Capacity to Co-exist with Natural Hazards and Their Consequences." In *Disaster Resilience: An Integrated Approach*, ed. D. Paton and D. Johnston (Springfield: Charles C Thomas Publisher, 2006): 3–10.

———. "Preparing for Natural Hazards: The Role of Community Trust." *Disaster Prevention and Management: An International Journal* 16, no. 3 (2007): 370–379.

Paton, D., L. Smith, and J. Violanti. "Disaster Response: Risk, Vulnerability and Resilience." *Disaster Prevention and Management* 9, no. 3 (2000): 173–179.

Paxson, C., E. Fussell, J. Rhodes, and M. Waters. "Five Years Later: Recovery from Post-traumatic Stress and Psychological Distress Among Low-income Mothers Affected by Hurricane Katrina." *Social Science and Medicine* 74, no. 2 (2012): 150–157.

PBS Religious & Ethics Newsweekly, "Hurricane Katrina Faith-Based Relief Effort." Last modified September 5, 2005, http://www.pbs.org/wnet/religionandethics/2005/09/02/september-2-2005-hurricane-katrina-faith-based-relief-efforts/12722/.

Peacock, W. G. "Hurricane Mitigation Status and Factors Influencing Mitigation Status Among Florida's Single-Family Homeowners." *Natural Hazards Review* 4, no. 3 (2003): 149–158.

Pelling, M. "Participation, Social Capital and Vulnerability to Urban Flooding in Guyana." *Journal of International Development* 10, no. 4 (1998): 469–486.

Pelling, M. and C. High. "Understanding Adaptation: What Can Social Capital Offer Assessments of Adaptive Capacity?" *Global Environmental Change* 15, no. 4 (2005): 308–319.

Pennington, M. *Robust Political Economy: Classical Liberalism and the Future of Public Policy* (Cheltenham: Edward Elgar, 2011).

Perelman, M. "Schumpeter, David Wells, and Creative Destruction." *The Journal of Economic Perspectives* 9, no. 3 (1995): 189–197.

Phillips, S. "What Went Wrong in Hurricane Crisis." *Dateline NBC*, interview transcript, 2005, http://thedemocraticdaily.com/2005/09/10/what-went-wrong-in-hurricane-crisis/.

Pike, J. "Spending Federal Disaster Aid Comparing the Process and Priorities in Louisiana and Mississippi in the Wake of Hurricanes Katrina and Rita." *Nelson A Rockefeller Institute of Government and the Public Affairs Research Council of Louisiana*, 2007.

Pipa, T. "Weathering the Storm: The Role of Local Nonprofits in the Hurricane Katrina Relief Effort." *Nonprofit Sector Research Fund* (Aspen Institute, 2006).

Pooley, J., L. Cohen, and M. O'Connor. "Links between Community and Individual Resilience: Evidence from Cyclone Affected Communities in North-West Australia." In *Disaster Resilience: An Integrated Approach*, ed. D. Paton and D. Johnston (Springfield: Charles C Thomas Publisher, 2006): 161–173.

Pope, J. "Katrina: One Year Later. East N.O. Priest Personifies Resilience." *Nola*. Last modified on September 03, 2006, http://www.nola.com/katrina/stories/index2.ssf?/katrina/stories/heroes_nguyen.html.

Portes, A. "Social Capital: Its Origins and Applications in Modern Sociology." *Annual Review of Sociology* 24 (1998): 1–24.

———. "Social Capital: Its Origins and Applications in Modern Sociology." In *Knowledge and Social Capital*, ed. E. L. Lesser (Boston: Butterworth-Heinemann, 2000): 43–67.

Poulsen, A. U. and G. T. Svendsen. "Social Capital and Endogenous Preferences." *Public Choice* 123, no. 1–2 (2005): 171–196.

Putnam, R. D. *Making Democracy Work: Civic Traditions in Modern Italy* (Princeton: Princeton University Press, 1993).

———. "Bowling Alone: America's Declining Social Capital." *Journal of Democracy* 6, no. 1 (1995): 65–78.

Quarantelli, E. L. *Disasters: Theory and Research* (Oaks: Sage, 1978).

Ray, S. and Cardozo, R. "Sensitivity and Creativity in Entrepreneurial Opportunity Recognition: A Framework for Empirical Investigation." *Sixth Global Entrepreneurship Research Conference* (London: Imperial College, 1996).

Riker, William H. *The Theory of Political Coalitions* (New Haven: Yale University Press, 1962).

———. *The Art of Political Manipulation* (New Haven: Yale University Press, 1986).

Richter, A. and J. Grasman. "The Transmission of Sustainable Harvesting Norms When Agents Are Conditionally Cooperative." *Ecological Economics* 93 (2013): 202–209.

Rodin, J. *The Resilience Dividend: Being Strong in a World where Things Go Wrong* (New York: Public Affairs, 2014).

Rodriguez, H., Trainor, J., and Quarantelli, E. "Rising to the Challenges of a Catastrophe: The Emergent and Prosocial Behavior Following Hurricane Katrina." *The Annals of the American Academy of Political and Social Science* 604, no. 1 (2006): 82–101.

Romer, D. *Advanced Macroeconomics*, 4th ed. (New York: McGraw-Hill Irwin, 2012).

Rose-Ackerman, S. "Altruism, Ideological Entrepreneurs and the Non-profit Firm." *Voluntas* 8 (1997): 120–134.

Rosegrant, S. "Wal-Mart's Response to Hurricane Katrina: Striving for a Public-Private Partnership." Kennedy School of Government Case Program C16–07–1876.0, *Case Studies in Public Policy and Management* (Cambridge, MA: Kennedy School of Government, 2007a).

———. "Wal-Mart's Response to Hurricane Katrina: Striving for a Public-Private Partnership (Sequel)." Kennedy School of Government Case Program C16–07–1876.1, *Case Studies in Public Policy and Management* (Cambridge, MA: Kennedy School of Government, 2007b).

Runst, P. "Post-Socialist Culture and Entrepreneurship." *American Journal of Economics and Sociology* 72, no. 3 (2013): 593–626.

Russell, G. "Chef Menteur Landfill Testing Called a Farce." *New Orleans Times-Picayune*, May 26, 2006, testimony of Reverend Nguyen; interview with James Bui, October 25, 2010.

Russell, S. "Voices from the Storm." *Jewish Action*, March 4, 2013, https://www.ou.org/jewish_action/03/2013/hurricane-sandy-a-spiritual-response/.

Sabetti, F. *Political Authority in a Sicilan Village* (New Brunswick: Rutgers University Press, 1984).

Sahlman, W. A. "Some Thoughts on Business Plans." In *The Entrepreneurial Venture*, ed. W. A. Sahlman, H. Stevenson, M. J. Roberts, and A. V. Bhide (Boston: Harvard Business School Press, 1996): 138–176.

Salamon, L. M., S. W. Sokolowski, and R. List. "Global Civil Society: An Overview." In *Global Civil Society: Dimensions of the Nonprofit Sector*, ed. L. M Salamon, S. W. Sokolowski, and Associates 2 (Bloomfield: Kumarian Press, 2003).

Sautet, F. *An Entrepreneurial Theory of the Firm* (New York: Routledge, 2000).

———. "Why Have Kiwis not Become Tigers? Reforms, Entrepreneurship and Economic Performance in New Zealand." In *Making Poor Nations Rich: Entrepreneurship and the Process of Development*, ed. Ben Powell (Redwood City: Stanford University Press, 2008): 364–396.

Schaeffer, E. C. and A. Kashdan. "Earth, Wind, and Fire! Federalism and Incentive in Natural Disaster Response." In *The Political Economy of Hurricane Katrina and Community Rebound*, ed. E. Chamlee Wright and V. H. Storr (Northampton: Edward Elgar, 2010): 159–182.

Schapiro, R. "Hurricane Sandy, One Year Later: At the Jersey Shore, a Slice of Heaven Devastated." *NY Daily News*. Last modified October 26, 2013, http://www.nydailynews.com/new-york/hurricane-sandy/sandy-1-year-jersey-shore-article-1.1493260.

Schelling, T. C. "Bargaining, Communication, and Limited War." *Conflict Resolution* 1, no. 1 (1957): 19–36.

Schelling, T. C. *The Strategy of Conflict* (Cambridge: Harvard University Press, 1960).

Schigoda, M. "Facts for Features: Hurricane Katrina Impact." News release, *Greater New Orleans Community Data Center*, 2011, http://www.datacenterresearch.org/data-resources/katrina/facts-for-impact/.

Schneider, M., P. E. Teske, and M. Mintrom. *Public Entrepreneurs: Agents for Change in American Government* (Princeton: Princeton University Press, 1995).

Schneider, S. "Who's to Blame? (Mis)perceptions of the Intergovernmental Response to Disasters." *Publius: The Journal of Federalism* 38, no. 4 (2008): 715–738.

Schreiner, M. "Scoring: The Next Breakthrough in Microcredit?" *Occasional Paper CGAP*, no. 7 (2003): 1–127.

Schuler, S. R., S. M. Hashemi, and H. Pandit. *Beyond Credit: SEWA's Approach to Women's Empowerment and Influence on Women's Reproductive Lives in Urban India* (Boston: JSI Research and Training Institute, 1995).

Schuler, S. R., S. M. Hashemi, A. P. Riley, and S. Akhter. "Credit Programs, Patriarchy and Men's Violence Against Women in Rural Bangladesh." *Social Science & Medicine* 43, no. 12 (1996): 1729–1742.

Schumpeter, J. *Capitalism, Socialism and Democracy* (London: George Allen & Unwin, [1942] 1976).

———. *The Theory of Economic Development: An Inquiry into Profits, Capital, Credit, Interest, and the Business Cycle* (New Brunswick: Transaction publishers, [1934] 2012).

Schutz, A. *The Structures of the Life-World* (Evanston: Northwestern University Press, 1973).

Seabright, P. "Managing Local Commons: Theoretical Issues in Incentive Design." *Journal of Economic Perspectives* 7, no. 4 (1993): 113–134.

Sewell Jr., W. H. "A Theory of Structure: Duality, Agency, and Transformation." *American Journal of Sociology* 98, no. 1 (1992): 1–29.

Shane, S. A. *A General Theory of Entrepreneurship: The Individual-Opportunity Nexus* (Northampton: Edward Elgar, 2000).

———. *General Theory of Entrepreneurship: New Horizons in Entrepreneurship* (Northampton: Edward Elgar, 2003).

Shane, S., E. A. Locke, and C. J. Collins. "Entrepreneurial Motivation." *Human Resource Management Review* 13, no. 2 (2003): 257–279.

Shane, S. and S. Venkataraman. "The Promise of Entrepreneurship as a Field of Research." *Academy of Management Review* 25, no. 1 (2000): 217–226.

Shaw, R. and K. Goda. "From Disaster to Sustainable Civil Society: The Kobe Experience." *Disasters* 28, no. 1 (2004): 16–40.

Shepherd, D. A. and N. F. Krueger. "An Intentions-based Model of Entrepreneurial Teams' Social Cognition." *Entrepreneurship Theory and Practice* 27, no. 2 (2002): 167–185.

Sherry, V. N. "$100 Million and Counting for Staten Island Buy-out Program." *Silive.com.* Last modified August 29, 2014, http://www.silive.com/eastshore/index.ssf/2014/08/buy-outs.html.

Shockley, G. E. and P. M. Frank. "Schumpeter, Kirzner, and the Field of Social Entrepreneurship." *Journal of Social Entrepreneurship* 2, no. 1 (2011): 6–26.

Shockley, G. E., R. R. Stough, K. E. Haynes, and P. M. Frank. "Toward a Theory of Public Sector Entrepreneurship." *International Journal of Entrepreneurship and Innovation Management* 6, no. 3 (2006): 205–223.

Shughart, W. F. "Disaster Relief as Bad Public Policy." *The Independent Review* 15, no. 4 (2011): 519–539.

Singleton, S. and M. Taylor. "Common Property, Collective Action and Community." *Journal of Theoretical Politics* 4, no. 3 (1992): 309–324.

Skarbek, D. B. "Occupational Licensing and Asymmetric Information: Post-Hurricane Evidence from Florida." *The Cato Journal* 28, no. 1 (2008): 71–80.

Skarbek, E. C. "Experts and Entrepreneurs." *Experts and Epistemic Monopolies* 17 (2012): 99.

———. "The Chicago Fire of 1871: A Bottom-up Approach to Disaster Relief." *Public Choice* 160, no. 1–2 (2014): 155–180.

Skarbek, E. C. and P. R. Green. "Associations and Order in the Cultural and Political Economy of Recovery." *Studies in Emergent Order* 4 (2011): 69–77.

Smith, A. *An Inquiry into the Nature and Causes of the Wealth of Nations* (Indianapolis: Liberty Fund, [1776] 1981).

Smith, D. J. and D. Sutter. "Response and Recovery after the Joplin Tornado: Lessons Applied and Lessons Learned." *The Independent Review* 18, no. 2 (2013): 165–188.

Smith, G. *Planning for Post-Disaster Recovery: A Review of the United States Disaster Assistance Framework* (Washington, DC: Island Press, 2012).

Sobel, R. and P. Leeson. "Government's Response to Hurricane Katrina: A Public Choice Analysis." *Public Choice* 127, no. 1–2 (2006): 55–73.

———. "The Use of Knowledge in Natural-Disaster Relief Management." *Independent Review-Oakland* 11, no. 4 (2007): 519–532.

Source for Learning News, "Board Member Doris Voitier Receives Profile in Courage Award." Last modified March 17, 2007, https://www.sourceforlearning.org/news.cfm?newsid=25.

Springer, C. G. "Emergency Managers as Change Agents." *Ideas from an Emerging Field: Teaching Emergency Management in Higher Education* 12, no. 1 (2009): 197–211.

———. "Achieving Community Preparedness Post-Katrina." *UNLV Institute for Security Studies Faculty Publications* (2011).

Sproule-Jones, M. *Governments at Work: Canadian Parliamentary Federalism and its Public Policy Effects* (Toronto: University of Toronto Press, 1993).

Storr, V. H. "All We've Learnt: Colonial Teachings and Caribbean Underdevelopment." *Journal Des Economistes et Des Etudes Humaines* 12, no. 4 (2002): 589–615.

———. *Enterprising Slaves & Master Pirates: Understanding Economic Life in the Bahamas* (New York: Peter Lang International Academic Publishers, 2004).

———. "Weber's Spirit of Capitalism and the Bahamas' Junkanoo Ethic." *The Review of Austrian Economics* 19, no. 4 (2006): 289–309.

———. "The Market as a Social Space: On the Meaningful Extra-Economic Conversations That Can Occur in Markets." *Review of Austrian Economics* 21, no. 2–3 (2008): 135–150.

———. "North's Underdeveloped Ideological Entrepreneur." In *Annual Proceedings of the Wealth and Well-being of Nations*, ed. E. Chamlee-Wright (Beloit: Beloit College Press, 2008–2009): 99–115.

———. *Understanding the Culture of Markets.* In Routledge Foundations of the Market Economy Series, ed. M. J Rizzo and L. White (New York: Routledge, 2013).

Storr, V. H. and A. John. "The Determinants of Entrepreneurial Alertness and the Characteristics of Successful Entrepreneurs." In *Annual Proceedings of the Wealth and Well-being of Nations*, ed. E. Chamlee-Wright (Beloit: Beloit College Press, 2010–2011): 87–107.

Storr, V. H. and S. Haeffele-Balch. "Post-Disaster Community Recovery in Heterogeneous, Loosely Connected Communities." *Review of Social Economy* 70, no. 3 (2012): 295–314.

Strömberg, D. "Natural Disasters, Economic Development, and Humanitarian Aid." *The Journal of Economic Perspectives* 21, no. 3 (2007): 199–222.

Sullivan Mort, G., J. Weerawardena, and K. Carnegie. "Social Entrepreneurship: Towards Conceptualization." *International Journal of Nonprofit and Voluntary Sector Marketing* 8, no. 1 (2003): 76–88.

Swedberg, R. "Social Entrepreneurship: The View of the Young Schumpeter." In *Entrepreneurship as Social Change: A Third New Movements in Entrepreneurship Book*, ed. C. Steyaert and D. Hjorth (Northampton: Edward Elgar, 2006), 21–34.

———. "Schumpeter's Full Model of Entrepreneurship: Economic, Non-economic and Social Entrepreneurship." In *An Introduction to Social Entrepreneurship: Voices, Preconditions, Contexts*, ed. Rafael Ziegler (Cheltenham, Northampton: Edward Elgar, 2009): 77–106.

Tedeschi, R. and L. Calhoun. "Posttraumatic Growth: Conceptual Foundations and Empirical Evidence." *Psychological Inquiry* 15, no. 1 (2004):1–18.

Thaler, R. and C. Sunstein. *Nudge: Improving Decisions about Health, Wealth and Happiness* (New York: Penguin Press, 2008).

Thompson, J. L. "The World of the Social Entrepreneur." *International Journal of Public Sector Management* 15, no. 5 (2002): 412–431.

Tierney, K. J. "Testimony on Needed Emergency Management Reforms." *Journal of Homeland Security and Emergency Management* 4, no. 3 (2007): 15.

Tierney, K., C. Bevc, and E. Kuligowski. "Metaphors Matter: Disaster Myths, Media Frames, and Their Consequences in Hurricane Katrina." *The Annals of the American Academy of Political and Social Science* 604, no. 1 (2006): 57–81.

Todd, H. "Women at the Center: Grameen Bank Borrowers after one Decade." *The Journal of Developing Areas* 31, no. 2 (1996): 276–279.

Tominc, P. and M. Rebernik. "Growth Aspirations and Cultural Support for Entrepreneurship: A Comparison of Post-Socialist Countries." *Small Business Economics* 28, no. 2/3 (2007): 239–255.

Torsvik, G. "Social Capital and Economic Development A Plea for the Mechanisms." *Rationality and Society* 12, no. 4 (2000): 451–476.

Trager, C. S. "Breezy Point Goes Its Own Way on Recovery." *Crain's New York Business.* Last modified 2013, http://www.crainsnewyork.com/article/2014/1020/REAL_ESTATE/310199993/breezy-point-goes-its-own-way-on-recovery.

Tullock, G. "The Politics of Bureaucracy." In *The Selected Works of Gordon Tullock, Volume 6: Bureacracy*, ed. C. Rowley (Indianapolis: Liberty Fund, [1965] 2005): 224–232.

Urban Land Institute, "New Orleans, Louisiana: A strategy for rebuilding." November 12–18, 2005, http://uli.org/wp-content/uploads/2012/11/NewOrleans-LA-05-v5.pdf.

Van Laerhoven, F. and C. Barnes. "Communities and Commons: The Role of Community Development Support in Sustaining the Commons." *Community Development Journal* 49, no. 1 (2014): 118–132.

Vien, N. "Written Testimony by Reverend Vien The Nguyen." *United States Committee on the Environment and Public Works*, February 26, 2007, http://webcache.googleusercontent.com/search?q=cache:8iYw6rAn_KoJ:www.epw.senate.gov/public/index.cfm%3FFuseAction%3DFiles.View%26FileStore_id%3D9f2dc64f-ad5d-4fc9–8673–467c63fc3a9a+&cd=1&hl=en&ct=clnk&gl=us.

Vollmer, H. *The Sociology of Disruption, Disaster and Social Change: Punctuated Cooperation* (Cambridge: Cambridge University Press, 2013).

Wacquant, L. J. "Negative Social Capital: State Breakdown and Social Destitution in America's Urban Core." *Netherlands Journal of Housing and the Built Environment* 13, no. 1 (1998): 25–40.

Wagner, R. E. "Pressure Groups and Political Entrepreneurs: A Review Article." *Papers on Non-Market Decision Making* 1, no. 1 (1966): 161–170.

Walker, B., S. Carpenter, J. Anderies, N. Abel, G. Cumming, M. Janssen, L. Lebel, J. Norberg, G. Peterson, and R. Pritchard. "Resilience Management in Social-Ecological Systems: A Working Hypothesis for a Participatory Approach." *Conservation Ecology* 6, no. 1 (2002): 14–31.

Walling, A. "The Katrina Success Story You Didn't Hear." *Regulation* 29, no. 1 (2006): 10–11.

Warner, C. "Census Tallies Katrina Changes, but the Changing New Orleans Area is a Moving Target." *Times-Picayune.* Last modified June 7, 2006, http://www.vendomeplace.org/press060706census.html.

Warner, C. and K. Darce. "Locals Not Waiting to Be Told What to Do." *The Times Picayune.* Last modified March, 12, 2006, http://www.vendomeplace.org/press031306neighborhoodefforts.html.

Weber, L. and Peek, L., eds. *Displaced: Life in the Katrina Diaspora* (Austin: University of Texas Press, 2012).

Weber, M. *Economy and Society: An Outline of Interpretive Sociology* (Berkley: University of California Press, 1978).

Weicher, J. C. "A Test of Jane Jacob's Theory of Successful Neighborhoods." *Journal of Regional Science* 13, no. 1 (1973): 29–40.

Weil, F. *The Rise of Community Engagement after Katrina* (New Orleans: Greater New Orleans Data Center and the Brookings Institution, 2010).

Wei-Skillern, J., J. E. Austin, H. Leonard, and H. Stevenson. *Entrepreneurship in the Social Sector* (Oaks: Sage Publications, 2007).

Wiggins, R. R. and T. W. Ruefli. "Schumpeter's Ghost: Is Hypercompetition Making the Best of Times Shorter?" *Strategic Management Journal* 26, no. 10 (2005): 887–911.

Winkler-Schmit, D. "Call to Action." *Times Picayune, New Orleans, Louisiana.* Last modified March 18, 2006, http://www.vendomeplace.org/press031806gambit broadmoor.html.

Woolcock, M. "Social Capital and Economic Development: Toward a Theoretical Synthesis and Policy Framework." *Theory and Society* 27, no. 2 (1998): 151–208.

———. "The Place of Social Capital in Understanding Social and Economic Outcomes." *Canadian Journal of Policy Research* 2, no. 1 (2001): 11–17.

Wright, J. D. *After The Clean-Up: Long Range Effects of Natural Disasters* (Beverly Hills: Sage Publications, 1979).

Yu, T. F. L. "Entrepreneurial Alertness and Discovery." *The Review of Austrian Economics* 14, no. 1 (2001): 47–63.

Yunus, M. "The Grameen Bank." *Scientific American* 281, no. 5 (1999): 114–119.

Zaheer, A. and S. Zaheer "Catching the Wave: Alertness, Responsiveness, and Market Influence in Global Electronic Networks." *Management Science* 43, no. 11 (1997): 1493–1509.

Zahra, S. A., E. Gedajlovic, D. O. Neubaum, and J. M. Shulman. "A Typology of Social Entrepreneurs: Motives, Search Processes and Ethical Challenges." *Journal of Business Venturing* 24, no. 5 (2009): 519–532.

Zakour, M. J. and D. F. Gillespie. "Effects of Organizational Type and Localism on Volunteerism and Resource Sharing during Disasters." *Nonprofit and Voluntary Sector Quarterly* 27, no. 1 (1998): 49–65.

Zhou, M. and C. Bankston. *Growing Up American: How Vietnamese Children Adapt to Life in the United States* (New York: Russell Sage Foundation, 1998).

Zwolinski, M. "The Ethics of Price Gouging." *Business Ethics Quarterly* 18, no. 3 (2008): 347–378.

Index

CPSIA information can be obtained
at www.ICGtesting.com
Printed in the USA
BVHW040311251119
564743BV00014B/426/P